CALIFORNIA CHANGING

50 SITES OF CLIMATE CHANGE IN AUGMENTED REALITY

BRETT SNYDER

ORO EDITIONS

Publishers of Architecture, Art, and Design
Gordon Goff: Publisher

www.oroeditions.com
info@oroeditions.com

Published by ORO Editions

Graphic Design: Cheng+Snyder
Text: Brett Snyder unless otherwise specified
ORO Managing Editor: Kirby Anderson

10 9 8 7 6 5 4 3 2 1 First Edition

Library of Congress data available upon request. World Rights: Available

ISBN: 978-1-961856-18-9

Color Separations and Printing: ORO Editions, Inc.
Printed in China.

International Distribution: www.oroeditions.com/distribution

ORO Editions makes a continuous effort to minimize the overall carbon footprint of its publications. As part of this goal, ORO Editions, in association with Global ReLeaf, arranges to plant trees to replace those used in the manufacturing of the paper produced for its books. Global ReLeaf is an international campaign run by American Forests, one of the world's oldest nonprofit conservation organizations. Global ReLeaf is American Forests' education and action program that helps individuals, organizations, agencies, and corporations improve the local and global environment by planting and caring for trees.

Introduction

Across the world, people look to California for inspiration. The state was one of the birthplaces of the environmental movement and it set the national standard for reducing auto emissions. It was the first US state to require solar panels on new homes. And it is where some of the newest technologies are being tested to mitigate climate change: from casting carbon in concrete to creating clouds that could, if the technology proves successful, lower the earth's temperature. California is the future.

But California also reveals a grimmer view of what is to come. In recent years, climate change has become increasingly tangible in the state. California is where rising sea levels and eroding cliffs intersect. It is where years of sustained drought converge with poor wildfire management, where the negative effects of the Gold Rush and decades of logging are still evident, where the harnessing of nature—through culverted rivers and dams—combined with the loss of Indigenous land practices has resulted in places that are less hospitable to people and animals. Finally, California is where the flow of water has enabled agricultural production while also contributing to the literal sinking of land in the Central Valley. Many of these scenarios raise questions about where and how development is done in sustainable ways.

This book looks at California through the lens of climate change on an architectural scale. It uncovers past visions of the future that would—if adopted—mitigate climate change. And it explores locations that are being adversely affected by climate change, such as historic sites that have been ravaged by wildfire, drought, or both. It looks at optimistic visions that dot the state like seeds for a sustainable future. And it describes community-led initiatives to remove public infrastructure that has negatively impacted habitats.

While a comprehensive accounting of all the places that contribute to, combat, or are affected by climate change is impossible, this book tries to make a broad swath of such places legible. By examining fifty sites across California—a number meant to suggest the fifty states that make up the United States and an acknowledgment of the many, often interwoven, factors surrounding climate change—the book offers a climate change tour of the state.

Each site is visualized through a text, an architectural illustration, and an augmented reality (AR) component. Though these modes are meant to complement one another, they can also be viewed independently.

I chose augmented reality to enhance the text and illustrations, because it is itself a metaphor for instability. The technology, which offers a temporal and spatial window into the material, can feel like magic. But there is no guarantee that the AR component will live as long as this book. In fact, it most likely will not. And just as AR enables a deeper read of the images, this book is intended to uncover a climate change reading of the places we experience every day. My hope is that, like the AR images, the text and illustrations will give access to new readings of the built environments that surround us and shape our everyday life.

How to Use this Book

This book is an atlas of sorts. It presents fifty sites or mini case studies that raise questions about how we live, what we value, and what issues we might consider as we plan for the future. To activate each site's AR content, look at the illustration through your phone with the Artivive app. The content revealed varies from image to image. Some tell the story of a site's past, others focus on the present. Taken together, all of the images weave a complex climate narrative suggestive of potential futures.

drawing by Maggie Ma and Aileen Louie

Key

Relationship to climate change:

- directly or indirectly contributes/contributed (for example, through greenhouse gas emission
- is/was directly affected (for example, by drought)
- is/was indirectly affected (for example, by an invasive species)
- is/was involved in mitigation efforts (for example, by lowering greenhouse gas emissions)

09 1 Aliso Canyon
11 2 Alta Wind Energy Center
13 3 Badwater Weather Station
15 4 Balboa Island House
17 5 Bateson Building
19 6 Capistrano Beach Boardwalk
21 7 Carlsbad Desalination Plant
23 8 Cherokee Studios
25 9 Church of One Tree
27 10 Corcoran Bowl
29 11 Donald C. Tillman Water Reclamation Plant
31 12 Dutch Slough
33 13 Fassler House
35 14 Foster City Levee
37 15 Fresno High-Speed Rail
39 16 Harris Ranch
41 17 Healdsburg Memorial Bridge
43 18 Heirloom Carbon Technologies
45 19 Inglewood Oil Field
47 20 Integral Urban House
49 21 Kaufmann Desert House
51 22 Kellogg House
53 23 Klamath River Dam Removal
55 24 Laguna Beach Goats
57 25 Lancaster Solar
59 26 Los Angeles River
61 27 Marine Cloud Brightening
63 28 Mighty Buildings
65 29 Millennium Tower
67 30 Mission Bay
69 31 Mission Creek Houseboat
71 32 NASA Ames Research Center
73 33 Oroville Dam
75 34 Pacifica Apartments
77 35 Parkside North Residence Hall
79 36 Platform Harmony
81 37 Salton Sea
83 38 San Francisco Federal Building
85 39 San Francisco Ferry Building
87 40 Sepulveda Adobe
89 41 Solar Star
91 42 Stanislaus Resource Recovery Facility
93 43 Stockton Super Commute
95 44 Sutter's Mill
97 45 The Theme Building
99 46 Tulare Lake Farms
101 47 Vantage Data Center
103 48 Whiskey Flat
105 49 Wilshire Grand Center
107 50 Yurok Tribal Lands

Contents

This book highlights fifty sites across California, arranged alphabetically. In the list to the left, sites are color coded based on their primary connection to climate change and plotted on the adjacent map. While this color indicates the main reason each site was selected, many sites intersect with multiple topics. As you peruse the book, you might find that other potential categorizations emerge (for example, positive/negative, geographic, and aspirational versus cautionary). Consider approaching these themes as prompts to look at our daily environment through the lens of climate change—not just in California but everywhere. Many of the included sites do not offer simple answers to mitigate climate change. Instead, they bring up questions of priorities (for example, whether to privilege restoration or development). As you view this material, consider the sites you know and what their relationship to climate change might be.

Aliso Canyon

Porter Ranch

The site of the largest methane leak in US history, Aliso Canyon Oil Field comprises several nondescript buildings, service roads, and pipelines in the Santa Susana Mountains in Los Angeles County.[1] In 1938, the Tidewater Associated Oil Company (owned by the billionaire oil industrialist J. Paul Getty) discovered oil 5,398 feet below ground and constructed dozens of oil wells. Much of this original infrastructure was repurposed in the 1970s for gas extraction. Eventually, the Aliso Canyon Natural Gas Storage Facility became the largest gas storage reservoir owned by the Southern California Gas Company (SoCalGas). It sits less than five miles from the terra-cotta-roofed ranch homes and landscaped cul-de-sacs of the Los Angeles suburb Porter Ranch. The field earned its infamous place in history in 2015 when it emitted one hundred thousand tons of methane over 112 days. During this time, residents experienced headaches, nosebleeds, and respiratory problems.[2] At its peak, the leak essentially doubled the amount of methane released from the entire Los Angeles basin.[3] According to the US Environmental Protection Agency, methane is almost thirty times as potent as carbon dioxide at trapping heat in the atmosphere. In the aftermath of the disaster, famed paralegal and environmental activist Erin Brockovich joined community members at the local high school to advocate on their behalf.[4] While the leak might seem like an aberration, reports suggest other failures by SoCalGas were kept confidential over the decades.[5] Though various groups have coalesced around calls to "shut it down," the Aliso Canyon oil facility has yet to cease operations.[6] What strategies might support better management of extractive sites like this that are so close to where people live?[7] The issue demands both short- and long-term solutions.

Alta Wind Energy Center

Tehachapi Pass

The Alta Wind Energy Center comprises about six hundred wind turbines, each approximately 420 feet tall, sitting on a total of 3,200 acres (about the same size as another energy-producing site, Culver City, well-known for its oil fields near residential neighborhoods).[8] Located in the Tehachapi Pass between Bakersfield and Los Angeles, the facility, also referred to as the Mojave Wind Farm, is the largest wind farm in the United States and the third-largest onshore wind farm in the world. Seen from the adjacent road, nestled in undulating hills, the turbines recede into the horizon in every direction—a sea of dispersed infrastructure among desert brush and Joshua trees. Collectively, the turbines have the capacity to produce 1,550 megawatts, enough to power about 450,000 homes. Wind power accounts for about 7 percent of California's total in-state energy generation. Obstacles to the widespread adoption of wind power have been numerous, including community resistance (especially when views are involved), the high cost of turbines, and the nature of wind power itself, which can fluctuate daily and seasonally.[9] But all technology has its challenges. According to a US Department of Energy report, wind could be a viable source of energy in all fifty states by 2050.[10] While most wind power is generated on large wind turbine farms like the Alta Wind Energy Center, the potential for smaller-scale wind power—such as the single turbine commonly used on agricultural farms—is significant.[11] Referred to as "distributed wind energy," these individual turbines can reduce energy costs and diversify power sources. As we phase out nonrenewables, at a variety of scales and situations, how can we take greater advantage of wind energy in all of its forms?

Badwater Weather Station

Death Valley National Park

In July 2021, the temperature reached 130 degrees Fahrenheit in Badwater Basin.[12] Previously, the hottest air temperature ever recorded on earth was 134 degrees in July 1913, also in Death Valley (though this higher record has been disputed over the years).[13] In September 2021, the National Park Service and National Weather Service announced a new, publicly accessible weather monitoring station in Badwater Basin, at the lowest point in North America, 219 feet below sea level.[14] The station resembles an Erector set, and its data is available online for anyone to monitor. A popular destination, it attracts scientists, selfie enthusiasts, and an emerging genre: extreme weather tourists. But why does it get so hot here? Death Valley is a long, narrow basin bounded by mountains—the Amargosa Range to the east, the Panamint Range to the west, the Sylvania Mountains to the north, and the Owlshead Mountains to the south. This unique geography forces hot air back down as it tries to rise, condensing it and trapping it in the depths of the valley.[15] As temperatures rise globally, places like Death Valley could become testing grounds for how people respond to extreme heat.[16] While more and more communities are grappling with rising temperatures, some are altering their lifestyles to avoid being outdoors during the hottest parts of the day. Although air conditioners have been used to create a sense of comfort in relatively tamer environments, the extremes in Death Valley can allow us to test what is tolerable and even adjust our expectations around climatic comfort. Perhaps by monitoring what we are capable of tolerating (here as well as in less extreme conditions), we might even find comfort outside the ranges that have come to be industry standards. How might society adapt to heat extremes without resorting to habits that continue to exacerbate the problem?

Balboa Island House

Newport Beach

Balboa Island, one of eight islands that are part of the city of Newport Beach, used to be not much more than a bean-shaped mudflat. But in the early 1900s, dredging from nearby channels was used to fortify the ground, making it possible for residential development. By the twenty-first century, it had developed into a wealthy, 130-acre harborside community.[17] Today, the island holds roughly 1,400 houses, all within walking distance of the coastline. The perimeter of Balboa Island is circumscribed by a paved concrete pedestrian boardwalk and piers that fan out like sunbursts. It boasts a main street with folksy shops that would have piqued Norman Rockwell's interest. Though the houses appear modest in scale, they average over $3 million in value. A popular destination for tourists, the island is accessible via bridge, ferry, and public docks. In 2018, the island community spent $1.8 million capping its seawall, originally built in 1920, with nine additional inches of concrete.[18] Most recent climate assessments predict an additional foot of sea level rise by the year 2100, which could translate to frequent flooding for Balboa Island, even with its upgraded barrier. For such communities—built in tidal zones and confronting rising sea levels—repairs are large in scale as well as cost. One proposed solution for Balboa Island is to raise the elevation of everything—houses, streets, sidewalks, utilities, and storm drains—so they are above the expected future flood levels. In hindsight, we can recognize the folly of building on such a sensitive and precarious coastal site, but what is the right approach for the future of such places—keep investing in them or try to remediate? Perhaps one way of absolving for the past is implementing plans that disincentivize any further development in this and other similar high-risk areas.[19]

drawing and AR by Bryan Vega

Bateson Building

Sacramento

Spanning an entire city block, the Bateson Building in Sacramento, named for the anthropologist Gregory Bateson, is a visionary public structure conceived by state architect Sim Van der Ryn, and is the first major building in the United States designed with operational energy reduction as a primary goal.[20] Completed in 1981, the unassuming building included several novel concepts to address energy usage when it originally opened. From the exterior, deep overhangs create shaded spaces for a passive cooling effect. On the inside, the building shows off its sustainable concepts both subtly and overtly. The interior is defined by a large atrium that was not conventionally air-conditioned. Louvers rotated to block light on south-facing clerestory windows to keep the sun out during peak daylight hours. Large canvas tubes circulated the atrium air with fans. On winter days, banners directed sunlight inside to help warm the space. Beneath the building, 660 tons of rock acted as a thermal mass, absorbing heat from building utilities during the day and warming the building at night. The air was moistened by an evaporative spray and fans. While many of these innovations ultimately proved unsuccessful or difficult to maintain, the building paved the way for a new generation of buildings that prioritize energy reduction. And despite some of its failures, the Bateson Building remains one of the state's most energy-efficient structures.[21] More than forty years later, many buildings still rely on energy-intensive methods for heating and cooling (think of typical modern glass office towers, for example, that would be heat-trapping greenhouses without mechanical cooling). Now, it is perhaps time to look again at many of the Bateson Building's passive techniques and adopt its lessons in sustainability and innovation more broadly.

Capistrano Beach Boardwalk

Dana Point

In 2023 and 2024, as part of a beach replenishment project, bulldozers spread sixty thousand cubic yards of sand from the Santa Ana River in Capistrano Beach. "Capo Beach," as it is known to locals, is at the southern end of Dana Point, about halfway between Laguna Beach and San Clemente, towns defined by their relationship to the Pacific Ocean. The slim strip of sand abuts a boardwalk, the tracks of Amtrak's Pacific Surfliner, the Pacific Coast Highway, and residential development on adjacent bluffs. Both the beach and these adjacent features are under threat from sea level rise, which is especially evident during seasonal storm surges.[22] Though sand levels rose and fell naturally in previous decades, ever since the 2015 El Niño, the beach has narrowed dramatically.[23] Storms have exposed protective measures from the past, including the fossils of old cars that were once part of a historic seawall. Other effects of erosion by tidal action have included disruptions to Amtrak, which is often out of service due to its tracks being covered with debris.[24] An active debate has developed over whether the tracks should be further protected from erosion or moved inland.[25] Measures such as adding riprap (large stones) and geotextiles have been tried over the years, but these have not been effective at preventing erosion. Organizations like the Surfrider Foundation have encouraged the California Coastal Commission to come up with a long-term solution to change the nature of the beach while preserving its public nature.[26] In a case like Capistrano Beach, where sand replenishment can be undone in a single storm, can we imagine more sustainable modes of beach revitalization?[27] The possibility exists, though it would involve removing concretized riverbeds, for example, which would allow such processes to happen naturally.

drawing by Langston Hay

Carlsbad Desalination Plant

Carlsbad

In a state chronically effected by drought, the prospect of turning seawater into drinking water might seem to offer a panacea. The Claude "Bud" Lewis Carlsbad Desalination Plant has been doing just that since 2015. Originally conceived in 1993 following several years of drought in Southern California, it is the largest seawater desalination plant in the United States, covering approximately six acres of public land in San Diego County. Though it took more than two decades to come to fruition, the plant now delivers about fifty million gallons of fresh water to San Diego County each day (enough for approximately four hundred thousand people) and accounts for about one-third of all water consumed in the county.[28] To process the seawater, suspended solids like sand are first removed. Then, in the reverse osmosis process, the seawater is forced through membranes that catch the salt. While removing salt from seawater might seem like a long-term solution to drought, the process is not without controversy.[29] The two most contentious factors are the high levels of energy required to process the seawater, and the resulting discharge of brine, which can adversely impact ocean life. The Sierra Club estimates that the daily energy required to run the Carlsbad plant is equivalent to the daily energy consumption of 6,200 households—a staggering amount.[30] And a 2019 report on the plant's environmental effects found mixed results. While a 2019 report on the plant's environmental effects found "no significant impact(s)" on sea life, the UC Santa Cruz researchers also noted that ocean salinity exceeded permitted levels, and had the potential to damage ocean ecosystems.[31] To date, several lawsuits have been brought against the plant, but none have been successful. If engineered solutions to climate change create new environmental problems, are they worth it?

Cherokee Studios

Los Angeles

What makes a building sustainable? Cherokee Studios is a 32,000-square-foot residential building in Los Angeles, completed by Pugh + Scarpa in 2009. It appears industrial thanks to its perforated metal facade. But behind its tough exterior are a series of sustainable features. These include solar panels, a green roof, and the facade itself, which allows its occupants to control cooling, shading, and views from inside by opening and closing large, perforated screens.[32] The double facade (the screen with windows and walls behind it) on the exterior is estimated to make the building nearly 50 percent more energy efficient than conventionally designed structures. From a heating and cooling perspective, windows are holes that waste energy. The double facade diminishes the energy loss by essentially adding insulation to the whole building. Cherokee was the first LEED (Leadership in Energy and Environmental Design) Platinum Certified building in Hollywood and the first LEED Platinum Certified multifamily building in Southern California.[33] Additional features include high-efficiency LED lighting, solar ventilation chimneys, sensor lighting, formaldehyde-free cabinetry, low volatile organic compound paints on the interiors, low-flow fixtures (plumbing that conserves water), and drought-tolerant native plants. A green roof captures 100 percent of stormwater runoff, and the roof's photovoltaic system generates thirty thousand kilowatt-hours. About 50 percent of the building's total energy use is from renewable sources.[34] Though the LEED system has been criticized for not focusing enough on building performance, the adoption of many of the building's features into California Code is a testament to the Cherokee's enduring success. Its sustainability concepts provide a palette of strategies for creative buildings that can have a positive impact, including ones that can be deployed to adapt underperforming existing homes.

drawing and AR by
Claire Adams and Kaitlyn Rivadeneira

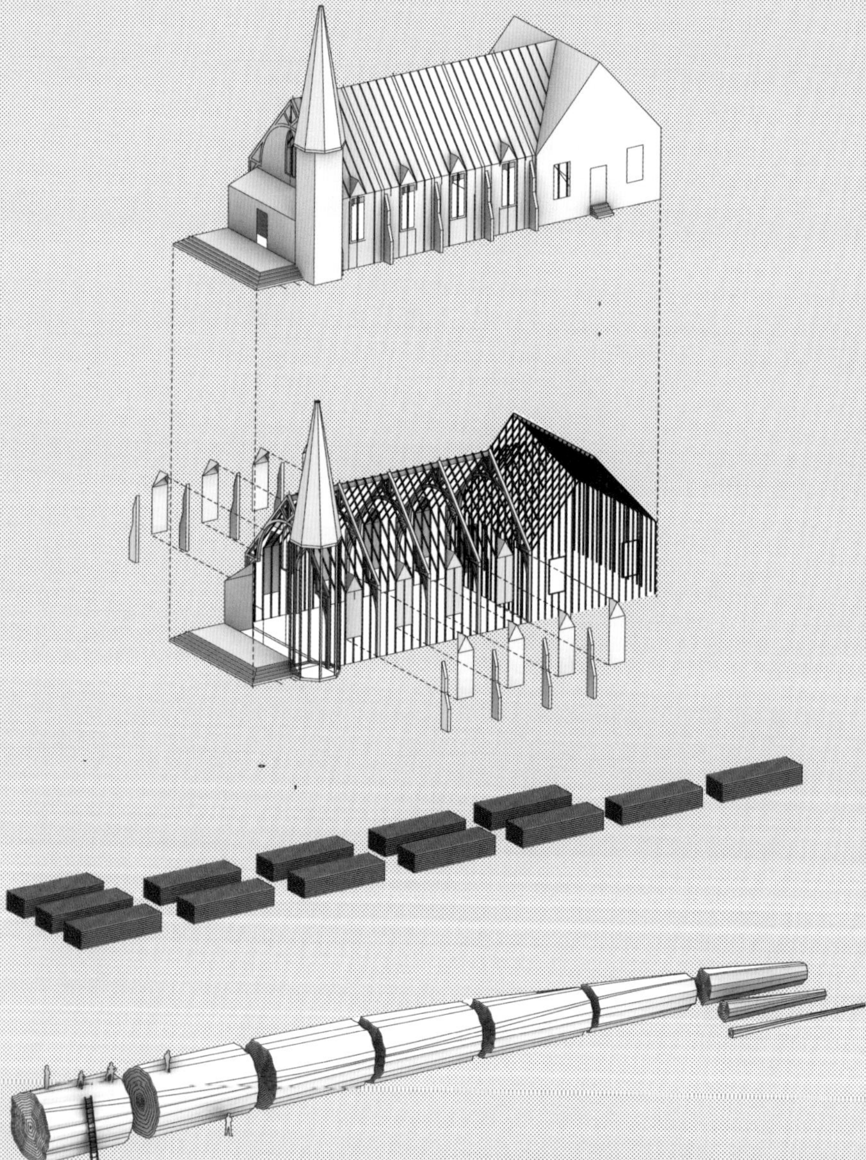

Church of One Tree

Santa Rosa

The Church of One Tree is a historic building constructed in the Gothic style and made almost entirely from a single old-growth redwood tree that was milled in nearby Guerneville, the site of an ancient forest. The tree used to make the church is estimated to be three thousand years old, 275 feet tall, and eighteen feet in diameter. Redwood was used for everything except the church's floor, which is made of pine. Originally completed in 1873, the building survived the 1906 earthquake (suffering damage only to its steeple), and was moved about half a mile to its current location in 1955.[35] Just as gold extraction on an industrial scale was harmful to the environment, so logging completely changed the nature of California's redwood forests. Though old-growth redwoods seemed like an endless resource in 1850—when they covered approximately two million acres—merely a century later only 4 percent of old-growth redwoods remained. The trees became the bridges, factories, railroads, and warehouses that make up California.[36] While the idea of creating an entire building from a single tree was a marketing strategy intended to highlight the utility of these giant trees, the actual fate of the redwoods was not so efficient. Logging practices in the 19th century were incredibly wasteful, with the Californian State Board estimating that as much as 72 percent of each tree was squandered.[37] This is even more tragic when one considers that a typical old-growth redwood can sequester around 250 tons of carbon (more typically sized trees might absorb just a single ton).[38] Only a short drive from the church is the two-thousand-year-old Clar Tree, which has persisted through centuries, evading loggers, drought, and even multiple lightning strikes.[39] How can we honor old-growth redwoods, which are not only a symbol of resilience but also remove more carbon than any other type of tree on earth?[40]

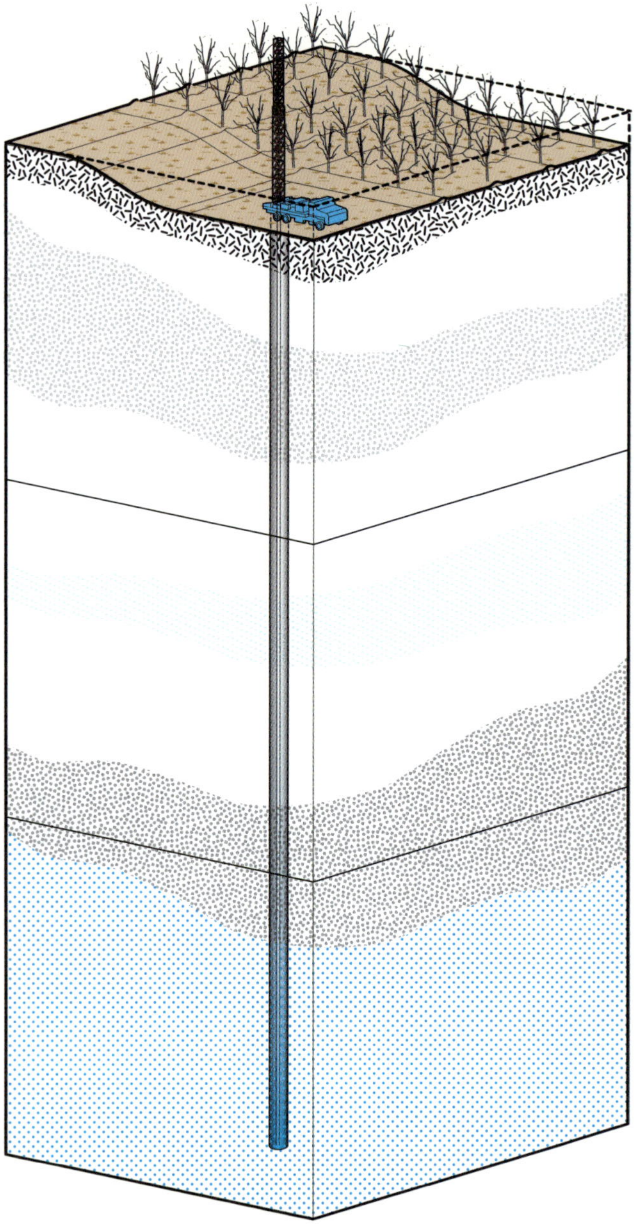

Corcoran Bowl

Corcoran

Located in the San Joaquin Valley is a vast area with a massive, slow-moving, yet nearly imperceptible problem. The town of Corcoran (population 21,960) and its surrounding land is sinking. Over the past fourteen years, the town has dropped as much as 11.5 feet in some places. That would put an entire single-story structure below ground. According to the US Geological Survey, this is one of the fastest-sinking areas in the country. But why? For decades, agricultural companies have used vast amounts of underground water to irrigate crops. As water is depleted from moist soil, the less dense dry soil begins to collapse in on itself, causing the ground to sink.[41] This process, known as subsidence, is one consequence of the way California manages water. In Corcoran, the use of groundwater has led to damaged or destroyed infrastructure and land that, paradoxically, is now more susceptible to flooding.[42] We must figure out how to manage water in a way that takes into account not just the maximization of agricultural production but also the conditions of the places where water is being depleted.[43] At the current pace, the Corcoran Bowl is estimated to sink another six to nineteen feet over the next two decades. But this problem is virtually invisible because it's spread out over such a vast area. While flood-level markers are typical along river courses, there are no equivalent public-facing measurements to make sinking land legible to community members. How might we address this issue? At the individual level, we might consider the water footprint of the food we eat. At a societal scale, we may need to rethink our food production from the ground up.

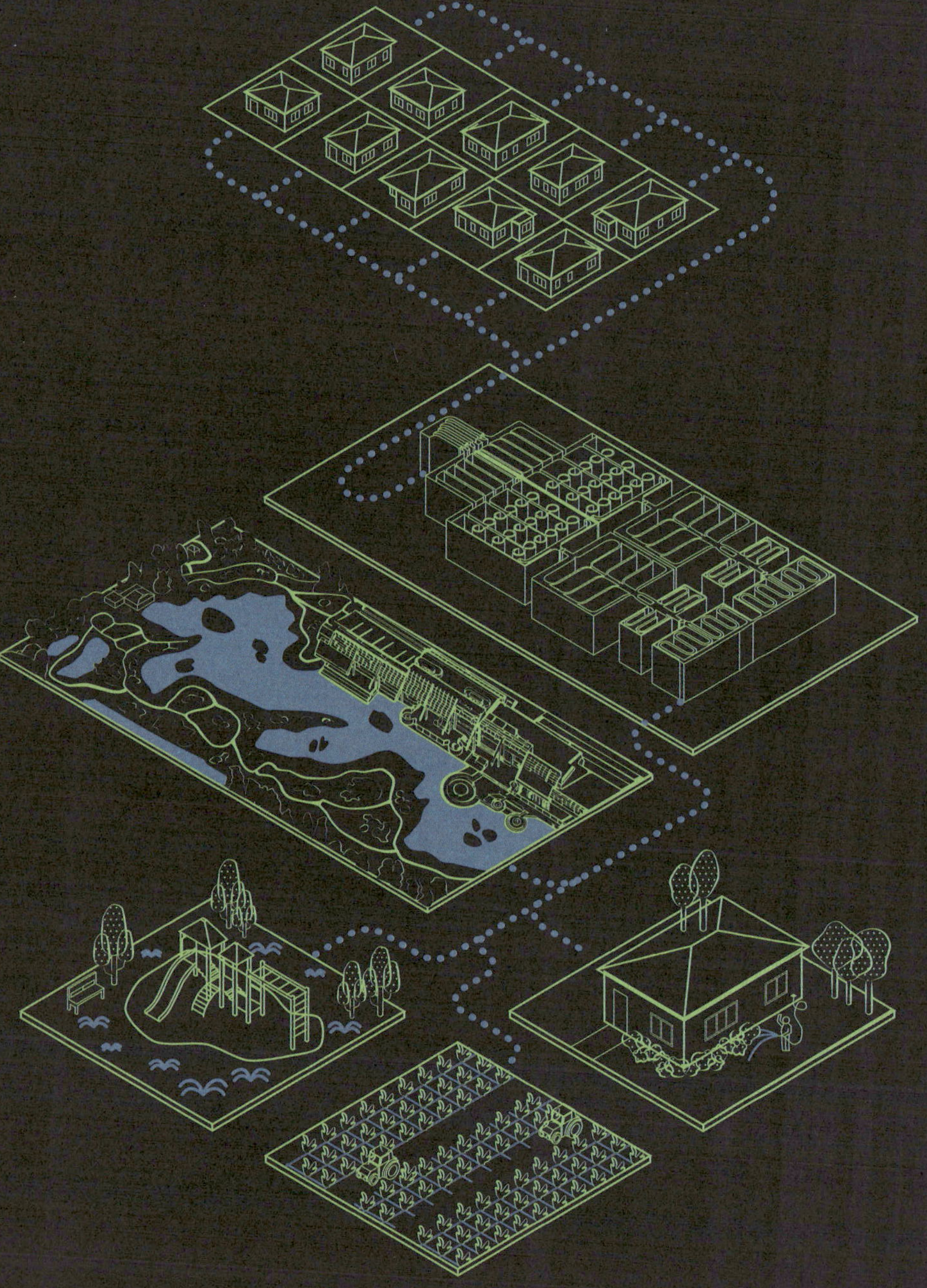

Donald C. Tillman Water Reclamation Plant
Van Nuys

The Donald C. Tillman Water Reclamation Plant and adjacent Suihō-en garden are designed to showcase the benefits that reclaimed water can have for people and the environment. About twelve hours before flowing through the Japanese garden, the water at Suihō-en is being flushed down the toilets of Los Angeles, en route to Tillman.[44] Located in the San Fernando Valley, the plant began operations in 1985 and each day processes eighty million gallons of wastewater, resulting in twenty-six million gallons of recycled water. The adjacent Suihō-en park is a 6.5- acre Japanese garden that includes a Zen meditation garden, a wet-strolling garden, and a tea garden.[45] The project embodies a simple but powerful architectural idea: Combine programs (waste collection, waste recycling, and a public amenity) to build awareness. Projects like Tillman—which is the only Japanese garden in the world directly connected to a water reclamation plant—might be considered a step toward normalizing the reuse of this precious resource by taking something viscerally unpleasant and transforming it into a place of beauty and delight. Already, this reclaimed water serves not just the adjacent garden (and more than twenty bird species) but also the golf courses and gardens of the greater Sepulveda Basin.[46] In 2023, the California State Water Resources Control Board unanimously voted to approve rules for recycling wastewater, commonly known as toilet-to-tap programs. As droughts increasingly impact California, how can we reduce stress on precious water sources (like the Colorado River) by tapping into the potential of recycled water even more?

drawing and AR by
Alex Motawi and Ariana Contreras

Dutch Slough

Oakley

Is flooding ever a good thing? In 2021, cheers rang out as an excavator removed a levee, allowing San Francisco Bay to expand, restoring a natural carbon sink, and answering that question with a hearty yes. A carbon sink is anything that absorbs more carbon than it releases. Located in the Sacramento–San Joaquin Delta, Dutch Slough is one of the most productive carbon sinks in the country. Over the last two hundred years, 90 percent of the bay's wetlands have been lost due to development and human activity. But according to UC Berkeley scientists, Dutch Slough ranks among the top 1 percent of natural carbon sinks studied worldwide, for the high rate at which it stores carbon.[47] The land was acquired for $28 million in 2003 by the California Department of Water Resources, and in 2018 the restoration began. Since then, 2.3 million cubic yards of dirt have been moved to raise the ground up to tidal marsh elevation (a specific height that allows it to be submerged or exposed depending on the level of the tides).[48] What makes it so efficient at storing carbon? As a restored wetland, regular tidal flooding helps to break down dead material, which then forms peat (soil made of partially decomposed organic matter), in which new plants are able to grow. The amount of carbon stored offsets the emissions produced by about one thousand cars annually.[49] In addition to carbon storage, Dutch Slough also provides habitat for wildlife, including salmon. Physically, the wetlands also act as a buffer to flooding for nearby communities such as Oakley. How can we appreciate and protect the seemingly inactive marshlands—unheralded areas that quietly provide enormous environmental benefits?

drawing by Stella Isaacs

Fassler House

Pacifica

As sea levels rise, houses precariously perched on the coast seem increasingly like disasters waiting to happen. Why aren't more of them deconstructed in advance of their inevitable demise? In 1939, Jean and Joseph Fassler purchased a single-room shack on the picturesque coastline of Pacifica, today a small city located between San Francisco and Half Moon Bay. The house was so close to the water, you could cast a line from its deck at high tide. Its beachfront view was shared with just a couple of other dwellings. Over the years, on weekends, Joseph expanded the house until it ultimately became a three-bedroom, two-bath residence with a two-car garage. As the Fasslers raised their family there, the house became a local landmark due to Jean's political career. When Pacifica incorporated in 1957, Jean became its first mayor, later serving on the San Mateo County Board of Supervisors, and she was also appointed to the Citizens' Advisory Committee on Recreation and Natural Beauty by the Johnson and Nixon administrations. Years after the Fasslers moved out, the home fell into disrepair under new owners.[50] The property, on one and one-third acres of land, was eventually purchased by the city, and the house was demolished as part of a "planned retreat"—that is, an organized effort to move structures out of harm's way in the face of climate dangers (in this case, rising sea levels).[51] Television crews watched as the house was torn down in minutes.[52] But why did the city remove the house in the first place? The removal was part of a larger restoration initiative that included widening the mouth of San Pedro Creek and providing spawning habitat for steelhead trout—changes that will ultimately help preserve the beach in ways the house itself hindered. While climate change was not a known issue when the house was built, its demolition raises a question: How do we decide what to maintain and what to tear down as sea levels rise?

Foster City Levee

Foster City

Can taller walls save us from rising sea levels? Just south of the San Mateo Bridge on the peninsula of the San Francisco Bay Area, a community is fighting future flooding with an approximately four-mile stretch of thick, plated steel. It begins seventeen feet below ground, where the steel is buried, before rising three-and-a-half feet above the existing twelve-foot levee. The exposed steel above the levee is capped with concrete and is expected to address a predicted eleven inches of sea level rise over the next twenty-five years.[53] Land behind the enhanced levee, formerly an area of tidal flows, was transformed into a residential development in 1962. To allow for this, eighteen million cubic yards of sand—equivalent to about thirteen Empire State Buildings—were dumped over former marshland.[54] Prior to the California Gold Rush, much of the edge of San Francisco Bay was tidal marsh. But an astonishing 90 percent of these marshes have now been lost to development. Zooming in, the area now known as Foster City historically allowed for tidal movement and provided a habitat for fish and birds. Today, however, this city of thirty-two thousand has a hard edge between homes and the bay. Although the steel levee is intended to keep out water until around 2050 (at a cost of around $90 million), some scientists think that hardening the edges of our coasts will just make flooding worse elsewhere. While taller walls are one response, a growing chorus of planners is proposing other ways of adapting. Ideas include restoring marshland (which can better absorb wave action than hard-edged walls), rewilding rivers that feed sediment to the bay (which can mean dismantling channels and dams), and even implementing planned retreat (which moves houses away from the edge of the bay).[55] As sea level rise becomes a reality, how communities respond will define the severity and complexity of the issue in the years to come.

drawing by Bhavi Patel

Fresno High-Speed Rail

Fresno

You cannot see it in Los Angeles or San Francisco yet. No tracks or stations are under construction. But in Fresno, the seeds of a transportation revolution are sprouting. In 2008, California voters approved a high-speed rail proposal that would connect Los Angeles and San Francisco by (a then-predicted) 2020. This 494-mile route currently has about 119 miles under construction (in the vicinity of Fresno and Bakersfield) and about 60 miles completed as of December 2024.[56] The construction sites offer a glimpse of the future rail infrastructure taking shape: smooth concrete bridges and parades of columns running through the back roads and fields of the Central Valley. While 2020 has come and gone without the finished product in sight, the vision for rail is more important than ever. The California High-Speed Rail Authority estimates that one round trip from San Francisco to Los Angeles by plane by plane emits a minimum of 390 pounds of carbon dioxide (CO_2), while one round trip by car emits at least 432 pounds of CO_2. In contrast, Amtrak routes are typically 34 percent more efficient than airplane travel and 46 percent more energy efficient than car travel over the same distance,[57] according to a 2021 US Department of Energy report.[58] The project has benefited from recent federal infrastructure grants. But the scope, scale, and ambition has led to numerous delays.[59] Some sections of track were built before preconstruction phases of the project were complete. Perhaps the biggest hurdle has been political—which has resulted in costly compromises such as a detour through the Mojave desert). Interested parties have arrived and left including bullet train companies from Europe and Japan that originally showed enthusiasm. With each passing year, the need for public infrastructure that lowers our collective carbon footprint increases. How can projects that are this big in scale—and importance—be better prioritized?

Harris Ranch

Coalinga

You can smell it from miles away. Its odor is familiar to anyone who has driven between San Francisco and Los Angeles. Located along Route 5 in the Central Valley is the Harris Ranch, California's largest beef producing company. Spread across nearly eight hundred acres, it is home to approximately one hundred thousand cattle on site at any time.[60] The scale of the operation is a testament to industrial agriculture's use of antibiotics to solve the problem of how to feed cattle in close proximity without spreading disease.[61] The ranch processes around 250,000 cattle annually, with sales in excess of $400 million. According to the US Environmental Protection Agency, a single cow can produce 150–265 pounds of methane per year.[62] Collectively, cows—through their belches and manure—are responsible for nearly 10 percent of all greenhouse gas emissions.[63] Methane traps more heat in the atmosphere per molecule than CO_2, and it can be more potent than CO_2 for the first twenty years after it is released, making it far worse for the environment in the short term.[64] This is consequential because of both the large carbon footprint associated with eating beef and the methane released during the life of the cow. As such, one part of the solution to global warming is reducing the amount of beef we consume. On average, Americans eat around fifty pounds of beef per year, but the good news is that beef consumption has been on a downward trend since the mid-1970s (and was surpassed by chicken consumption in 2010). Critics often downplay the potential of personal lifestyle choices to help mitigate climate change, but given the scale of the aggregate effects of cultural preferences, such dietary choices could have significant impact. Perhaps, at a minimum, anyone that chooses to eat meat should visit Harris Ranch to experience the scent of our collective consumption.

Healdsburg Memorial Bridge
Healdsburg

As temperatures change, some invasive species begin to thrive. One example is the plant known as arundo (*Arundo donax*), a giant reed native to East Asia that now populates the banks of the Russian River, including areas around the Healdsburg Memorial Bridge. Arundo was first planted in California in the 1820s by Spanish settlers. While its initial spread was due to its popularity as a roofing material (and for items like walking sticks and fishing poles), it has also been grown to stabilize the banks of rivers.[65] But this use has negatively affected existing ecosystems. According to the UC Riverside Center for Invasive Species Research, arundo can grow as much as four inches per day and reach a height of approximately twenty-five feet. It uses about three times as much water as native species and can become so dense that it kills off native plants. Flooding can uproot large rafts of arundo, which then damage the bases of bridges as they float downstream.[66] In California, arundo can be particularly problematic because it provides fuel for fires in already sensitive environments suffering from drought and a lack of prescribed burns.[67] Several counties have formed task forces to manage the spread of arundo. These groups promote its removal and discuss strategies for managing land in ways that are more sustainable for people, plants, and animals.[68] Although not all non-native species harm habitats, the damage arundo has caused along the Russian River prompts a question relevant to everyone in California and beyond: What steps can we take both to remove invasive species that damage existing ecosystems and promote plants that support resilient landscapes?

drawing by Bhavi Patel

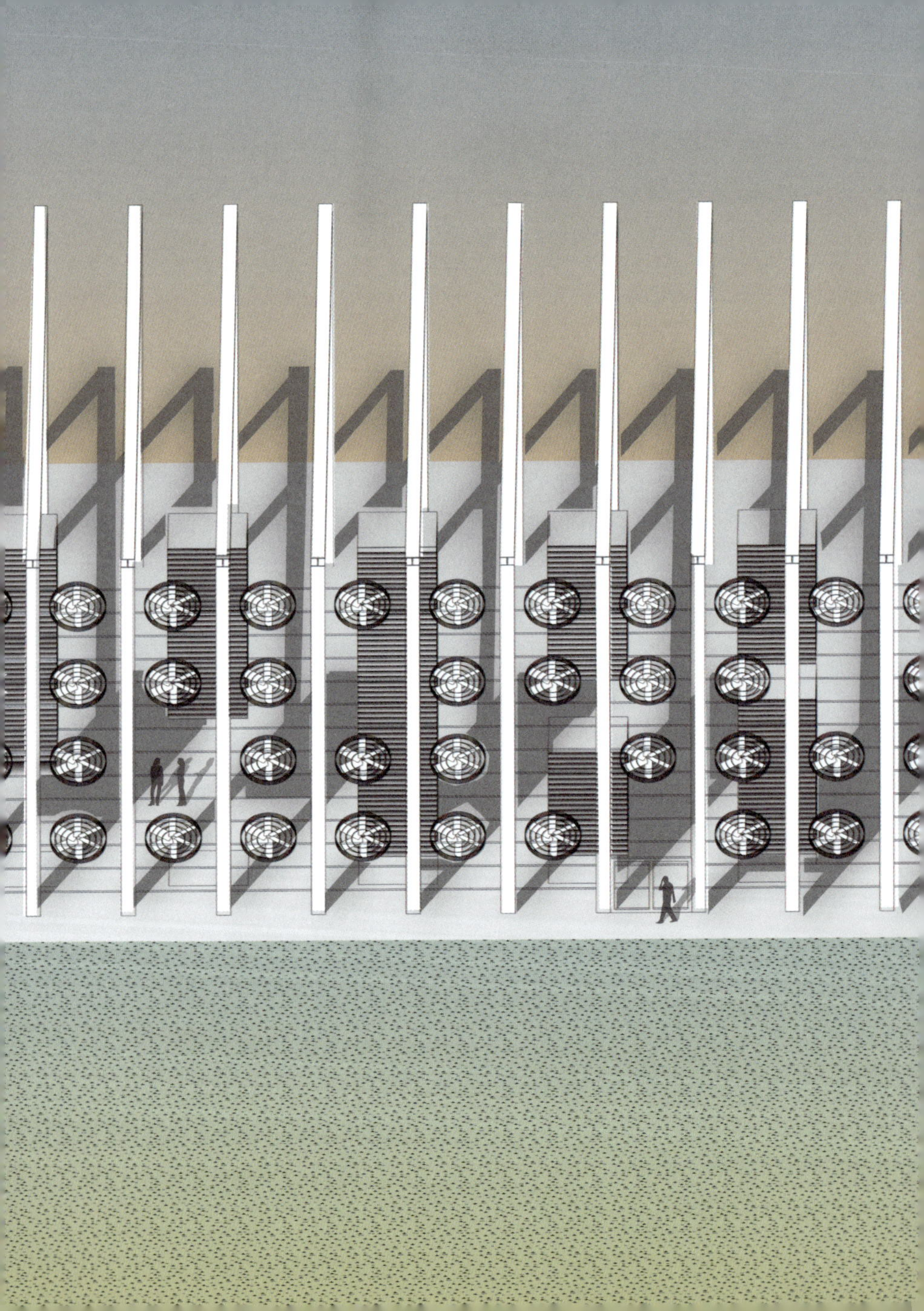

Heirloom Carbon Technologies
Tracy

What if we could suck carbon right out of the air? A facility in Tracy, a small city about sixty miles east of San Francisco, is doing just that. Bearing some resemblance to a giant air conditioner, it is the first direct carbon capture operation in the nation. Just as forests act as carbon sinks by removing carbon from the air, the start-up behind the Tracy facility—Heirloom Carbon Technologies—is vacuuming carbon from the air with a gridded arrangement of fans. The carbon is absorbed on a series of trays that are arranged in vertical columns. On each tray sits a powdery white substance that turns crusty once the carbon has been absorbed. The carbon is then permanently embedded into concrete so it does not have a warming effect on the planet.[69] The company will make money by selling carbon credits (which it has already sold to companies like Microsoft). Currently, Heirloom is able to capture approximately enough carbon to offset the emissions of two hundred cars per year—but it has plans to expand. Critics of the process say it will slow the transition to renewables and that it will never have the capacity to absorb enough carbon to have a measurable effect on climate change. This and technologies like it are also still expensive.[70] Nevertheless, Heirloom is not alone. A similar facility operates in Iceland, and hundreds of companies globally are developing carbon capture technologies. While direct carbon capture is not a single solution to climate change, could it be part of a larger toolbox of techniques?[71]

Inglewood Oil Field

Inglewood

From the sky, the paths around them look like a network of amoebas—a maze of dirt paths that each end in an amorphous bulge. From the pedestrian perspective, the oil derricks in Inglewood cast a striking yet anachronistic silhouette along the horizon of Baldwin Hills and Culver City on the west side of Los Angeles. While the sight of oil pumps may seem at odds with the glamour of Los Angeles and its identity as a hub for film and television, the oil boom has also played a key role in the city's development. What makes the image of the pumps particularly disturbing is their proximity to homes, schools, and sports fields. Many of these locations are separated from the oil pumps by just a chain-link fence. Covering nearly one thousand acres, the Inglewood Oil Field is one of the nation's largest. It was discovered in 1924, when California was a hotbed of oil exploration. In the decade since 2012, the site has produced 2.5–3 million barrels of oil each year and has yielded nearly 400 million barrels since it began operating.[72] Each barrel of oil contributes incrementally to climate change through its extraction, production, transport, and burning. Those who live near the oil fields experience heightened health risks, including for asthma, birth defects, and cancer.[73] An estimated thirty million barrels of oil remain in the ground. Though phase-out plans are already in place for this location—thanks to the years-long efforts of community members and environmental activists—what would it take to stop the extraction process now?[74] Is the sight of children playing baseball with oil pumps in the background not a jarring enough image?

drawing by Bhavi Patel

Integral Urban House

Berkeley

On an eclectic block of Victorians, Craftsman homes, and industrial buildings sits what was once a visionary architectural experiment. The experiment occurred at 1516 5th Street in Berkeley from 1974 to 1984. Save for its idiosyncratic side extension, you would hardly recognize the house from the famous book cover emblazoned with its image. The project was the brainchild of a group of academics—including Sim Van der Ryn, who became well-known as the California state architect under Governor Jerry Brown, and Bill and Helga Olkowski, authors of the *City People's Book of Raising Food* (1975). Collectively, they hoped to demonstrate that by changing our daily practices we could reduce the ecological footprints of our homes.[75] It was a radical idea at the time: Take a typical house and literally and metaphorically rewire it for sustainability. It would be solar powered and include crops to feed animals. In the words of Bill and Helga, the house and houses like it would help "make cities ecologically stable and healthy places to live."[76] The Integral Urban House was not just a house but a social experiment that was documented in *The Integral Urban House: Self-Reliant Living in the City*, published in 1979. This accompanying book explains urban homesteading concepts such as the negative consequences of a lawn (which results in labor, pesticides, and fuel consumption) versus an alfalfa patch (which can feed rabbits, creating rabbit meat, and natural compost).[77] Though some considered the project a failed experiment, today its influence can be seen not only through the widespread adoption of solar power but also through practices such as building performance evaluations, which aim to understand the efficiency of buildings as they are being used.[78] Can a reexamination of this project change our approach to architecture, ensuring we look at buildings not just as an assembly of material choices but also as a critical enactment of everyday lifestyle choices?

drawing and AR by
Joel Kjellberg and Danae Navarrette

Kaufmann Desert House

Palm Springs

The Kaufmann Desert House is an icon of modern architecture designed by Richard Neutra and built in 1946 (the name Kaufmann is familiar to many architects as the client who commissioned Frank Lloyd Wright's Fallingwater).[79] The Kaufmann house set off an architectural trend that has made Palm Springs a destination for design enthusiasts.[80] The house is a dynamic composition of rectangular planes made of concrete, stone, and glass that erupt from the rocky landscape. It was designed in response to the arid environment, with shaded breezeways strategically located to keep the house cool during the day and stone areas that emit heat during the cool nights. But the desert environment is transforming due to climate change, and passive techniques are no match for rising temperatures. Though the house originally included radiant heating and cooling, rooftop air-conditioning units were added as discreetly as possible in the 1960s.[81] In 2024, Palm Springs experienced eighty-three days above 110 degrees Fahrenheit (compared to twenty days in 1946). The number of extreme heat days (defined as above 115 degrees) in Palm Springs is predicted to increase to twenty-eight by mid-century and to fifty by the end of the century.[82] In July 2024, a new heat record was set: 124 degrees Fahrenheit.[83] Places like Palm Springs are at the forefront of climate change and an ideal testing ground for extreme heat. Residents are acutely aware of heat-related phenomena. Speaking with *The New York Times* about living in Palm Springs, resident Maggie Miles said, "My debit card melted in my drink holder. I've seen the internal car temperature get up to 150 degrees."[84] Could the Kaufmann House, once an icon of pleasant indoor-outdoor living and harmony between humans and nature, become a symbol of our shift toward a more discordant relationship with an increasingly inhospitable environment?

drawing and AR by
Tabitha Saunders and Liang Qiao

Kellogg House

Rich Bar

Gold was first found along the north fork of the Feather River in July 1850. Miners were drawn to the town of Rich Bar by stories of a lake lined with gold, told by the prospector Thomas Stoddard of Pennsylvania. The town was further popularized through what became known as the "Shirley letters," correspondence between "Dame Shirley" (Louise Amelia Knapp Smith Clappe) and her sister in Massachusetts. The letters were reprinted between 1854 and 1855 by *The Pioneer* magazine, which depicted local mining life, including its dangers, and the tensions between pioneers and Native Americans in the early 1850s.[85] Rich Bar built up quickly to accommodate the influx of miners, but the boom years were brief, lasting only until 1853. Still, the Gold Rush town and the surrounding Plumas County helped build California's population, putting it on the path to statehood in 1890. Over a century later, what had become a ghost town, its Gold Rush days a distant memory, was virtually wiped from the map by the Dixie Fire, which began in Feather River Canyon on July 13, 2021, and burned 963,309 acres. The Gold Rush–era Kellogg House was one of the 1,329 buildings destroyed.[86] It was built in 1852 from lumber sourced from the Rich Bar sawmill. Notably fires like the Dixie Fire exemplify the confluence of two factors: years-long drought due to climate change and the absence of prescribed burnings, a hallmark of Indigenous land management. Speaking with the *Los Angeles Times,* Shelby Leung, a local firefighter, explained that his elders from the Maidu tribe "would just burn every season—every fall and in the springtime . . . it was, like, patchwork where they would burn around their homes. It was just clearing that made it safer."[87] Additional impacts of the Gold Rush included the contamination of rivers and depletion of old-growth forests. How can we care for California's historical sites in a way that moves beyond a boom-and-bust cycle to one of restoration and repair?[88]

Klamath River Dam Removal

Hornbrook

Dams are a key tool that humans have used to harness nature. The construction of large-scale hydroelectric dams was particularly pronounced from the 1930s to the 1960s. As some of these large engineering projects begin to require repairs or face demolition, however, we are forced to grapple with their legacy.[89] The Klamath River dam removal project, the largest of its kind in history, is part of a new wave of projects that are dismantling infrastructure as communities begin to understand the negative consequences dams can have on the environment. The Lower Klamath Project is a network of four dams built from 1911 to 1962 that provided hydroelectric power but also changed the nature of the river, inhibiting sediment flow, removing habitat for native species (like salmon), and introducing non-native fish such as perch, catfish, and bass. Years of protests from Indigenous tribes and environmental activists resulted in a plan to remove the dams and let the river assume its historic, winding path.[90] Speaking with the *Los Angeles Times,* Leaf Hillman, a member of the Karuk tribe said, "It feels like a cleansing that is long overdue. This river has been literally dying for years, having the life strangled out of it by these dams. . . . I love seeing that sediment being pushed down, because that's a river acting like a river."[91] While many projects of the twentieth century were major engineering feats (including highways, dams, and systems like the California State Water Project), the most important engineering feats of our current century could be the strategic removal of this infrastructure and to let nature heal itself. As we recognize the importance of natural processes for buffering the effects of climate change, how might we evaluate our legacy infrastructure to determine what would best be unbuilt?

Laguna Beach Goats

Laguna Beach

The stunning views of the coast are accompanied by rustling grass, buzzing flies, and *baas*. Look closer and the source of that last sound becomes clear: a gathering of two hundred goats, one of two herds currently patrolling Laguna Beach.[92] Nearby, several homes have a front-row seat to the action. The goats are spread out across the rugged terrain, some in groups and some right up against an electric fence that keeps them contained. If you live in California, you might be used to a similar sight: goats grazing on a highway shoulder or on grassy hills. Why all the goats? Their job, overseen by trained shepherds, is to eat the grass and reduce fuel for fires.[93] While California is naturally home to seasonal wildfires, they have grown worse due to higher temperatures and many years of drought. The use of goats to reduce fuel is admittedly low tech. Inspired by an approach launched in Oakland, following the Oakland Hills Fire of 1991, Laguna Beach adopted its own goat program. Speaking with ABC News, the fire marshal of Laguna Beach, James Brown, explained, "The City of Laguna Beach has been using goats . . . for over twenty-five years, and they have proven very effective at reducing the dead vegetation fuel load and helping to protect our City. . . . They have also been very cost-effective."[94] After the 1993 wildfire in Laguna Beach, observers noted that in an area where a small herd of goats had been grazing, no houses were lost; in other areas, four hundred homes were destroyed.[95] In 2024, the two herds of goats were able to trim approximately 280 acres. Recruiting goats for ecosystem maintenance leads us to ask what other creative, low-tech strategies might be devised to reduce natural disasters?

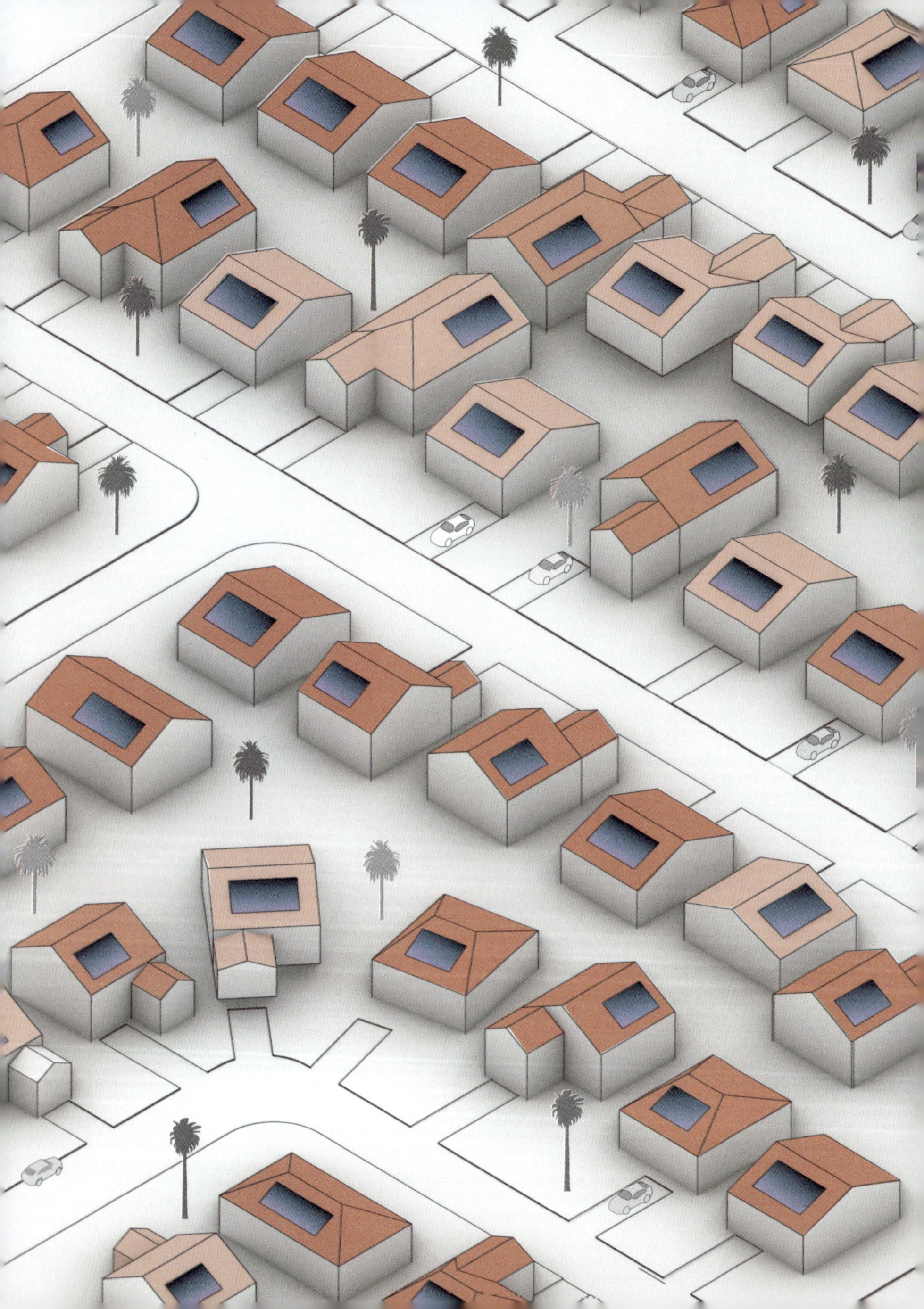

Lancaster Solar

Lancaster

In 2013, Lancaster was at the forefront of how Californians would get their power in the future. That year, Lancaster, a charter city in northern Los Angeles County (population 169,000) became the first city in California to require solar panels in new construction. Speaking to *The New York Times* when Lancaster was in the process of passing this legislation, Mayor R. Rex Parris said, "We want to be the first city that produces more electricity from solar energy than we consume on a daily basis."[96] In his quest to make Lancaster the center of solar innovation, Parris pushed many of the city's major buildings to incorporate solar power, including City Hall, a performing arts center, and a stadium, which collectively generate 1.5 megawatts of electricity (about enough to power 1,300 homes).[97] Other photovoltaic arrays can be seen on churches, schools, medical offices, car dealerships, and even the Los Angeles County State Prison (which has a 2,634 kilowatt photovoltaic system).[98] And just as Lancaster helped raise the solar bar, California has led the nation in the solar transition. California homes produced more solar energy in 2022 than any other state, resulting in twenty-three million megawatt-hours of power.[99] Through the leadership provided by communities like Lancaster, as of January 1, 2020, California began to mandate solar photovoltaics for all new construction.[100] The requirement includes single-family homes, condominiums, and apartment buildings less than three stories high, as well as most nonresidential construction. Could this small and sunny town's bottom-up innovation demonstrate that local communities and governments have the power to drive widespread adoption and change in the transition to clean energy?[101]

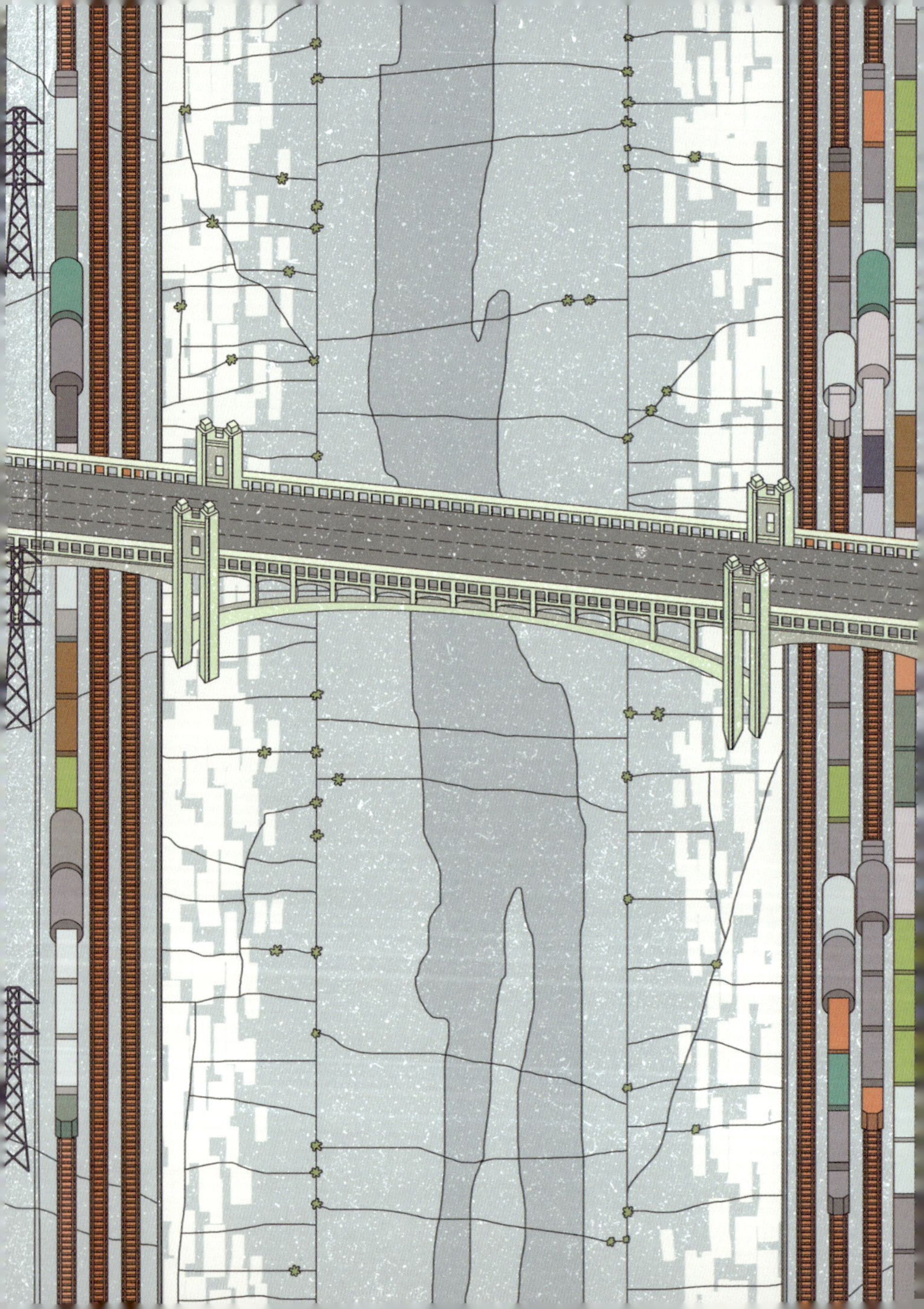

Los Angeles River

Los Angeles

Originally called the Paayme Paxaayt (West River) by the Tongva people, the Los Angeles River watershed covers 834 square miles, about twice the area of the city of Los Angeles itself. It flows from its headwaters in the Simi Hills and Santa Susana Mountains to its mouth in Long Beach, where it empties into San Pedro Bay. Along this fifty one-mile path, 95 percent of the riverbed—which is up to six hundred feet wide in some places—has been replaced by concrete.[102] The channelization, completed by the US Army Corps of Engineers—which, as of 2024, still shares oversight of the waterway—consists of a concrete basin and embankments to prevent flooding (the project was initiated in response to major flooding in 1938 that killed eighty-seven people).[103] Today, this channel looks less like a river and more like a multilane freeway, sprinkled with the occasional shopping cart and at some points running straight as far as the eye can see. While this concretized path quickly moves water from the mountains to the sea, it also contributes to polluted runoff, harming adjacent communities as well as fish and wildlife habitats. Over the last forty years, activists, environmentalists, and Indigenous communities have pushed for the rewilding of the river. The groups have held events such as an annual cleanup known as the "Gran Limpieza" (Great Cleanup), which draws up to 2,500 people each spring.[104] Responding to and building on this advocacy work, architects and planners are preparing a major river restoration and revitalization.[105] Achieving their goals will require compromises. Once people live on a river, it cannot meander as it once did. Allowing greater width in one place would undoubtedly impact another (by creating a greater flood risk, for example). Perhaps the memory of a river that was once a source of life to steelhead trout, the North American beaver, and many other species should be a guiding template. A restoration plan that honors both the social and environmental histories would undoubtedly be the most resilient to the climate challenges ahead.

drawing by Stella Isaacs

Marine Cloud Brightening

Alameda

It sounds like a science fiction plotline: engineered clouds to cool the planet and prevent disaster. But the first US-based outdoor experiments to test this idea took place in 2024 on a former military pier turned museum in Alameda (an island community in the San Francisco Bay itself at risk from sea level rise).[106] How did this effort begin? While on a hike with his son in Wales, the British physicist John Latham paused to look at the clouds. His son asked why they were shiny on top and black on the bottom. Latham explained that the clouds were mirrors to the sun, and that "aha" moment birthed the Marine Cloud Brightening project.[107] Fast-forward some thirty-plus years to a bright Tuesday in April 2024: The deck of the USS *Hornet* became the first testing ground for this novel idea, though the experiments were soon paused when Alameda officials expressed safety concerns.[108] To make the clouds brighter, scientists sprayed microscopic sea salt into the air to boost their reflectivity. The clouds in the Bay Area tend to be low and thin in spring and summer, perfect conditions for the experiment. Speaking with the *San Francisco Chronicle,* Rob Wood, a scientist at the University of Washington remarked that, if 15 percent of the earth's marine clouds were brightened, the planet would cool by roughly one degree Fahrenheit.[109] While we have been told for years that legislation or behavioral change are the main answers to ameliorating climate change to reduce carbon emissions, neither has been enough. Thirty years ago fossil fuels provided around 80 percent of the world's energy and today that percentage remains stubbornly the same.[110] Perhaps, more than fifty years after the first United Nations Earth Summit in Rio de Janeiro, it is time to test even ideas that seem a bit hazy. At the same time, it is important to remember that technological quick fixes have often led to unanticipated and unwanted consequences.

drawing by Meloddy Robles

Light Stone
3D Material
Steel Frame
PU Foam
Gypsum Board

Mighty Buildings
Desert Hot Springs

While many people think transportation is the leading cause of greenhouse gasses, the building industry—through a combination of materials and operations—is actually the largest contributor, with around 40 percent of total CO_2 emissions compared with transportation's 25 percent.[111] What if buildings could have a smaller environmental footprint in terms of both embodied and operational energy?[112] One company based out of Oakland is aiming to build net-zero energy homes (homes that produce as much energy as they use) by using 3D printing to reduce construction waste by up to 90 percent.[113] The company, Mighty Buildings, completed its first prototype in the cactus-filled town of Desert Hot Springs in 2022. Designed by EYRC Architects, this home evokes the pragmatism of designer John Entenza's postwar Case Study House Program.[114] And it has similar ambitions: to produce an affordable house for the masses and to change how we build such houses. Currently, net-zero energy homes are only a small proportion of homes built. In 2021, a nonprofit counted just 24,500 such homes in the United States.[115] With the growing popularity of rooftop solar, however, this number is sure to rise. But a house's energy footprint also includes the materials that went into its construction, as well as what happens to those materials at the end of their life cycles.[116] The potential benefits of 3D printing include reducing construction waste (because 3D printing can be done to exact dimensions without waste). Other advantages include using low-carbon-footprint materials (some designers are experimenting with printing mud, for example), and more directly connecting architects to the process of building (in other words, leapfrogging a construction industry that has been slow to change).[117] It remains to be seen whether 3D printing buildings can offer a truly viable and scalable approach to producing more sustainable buildings and cities.

drawing and AR by
Cameron Gillern and Jaime Jeong

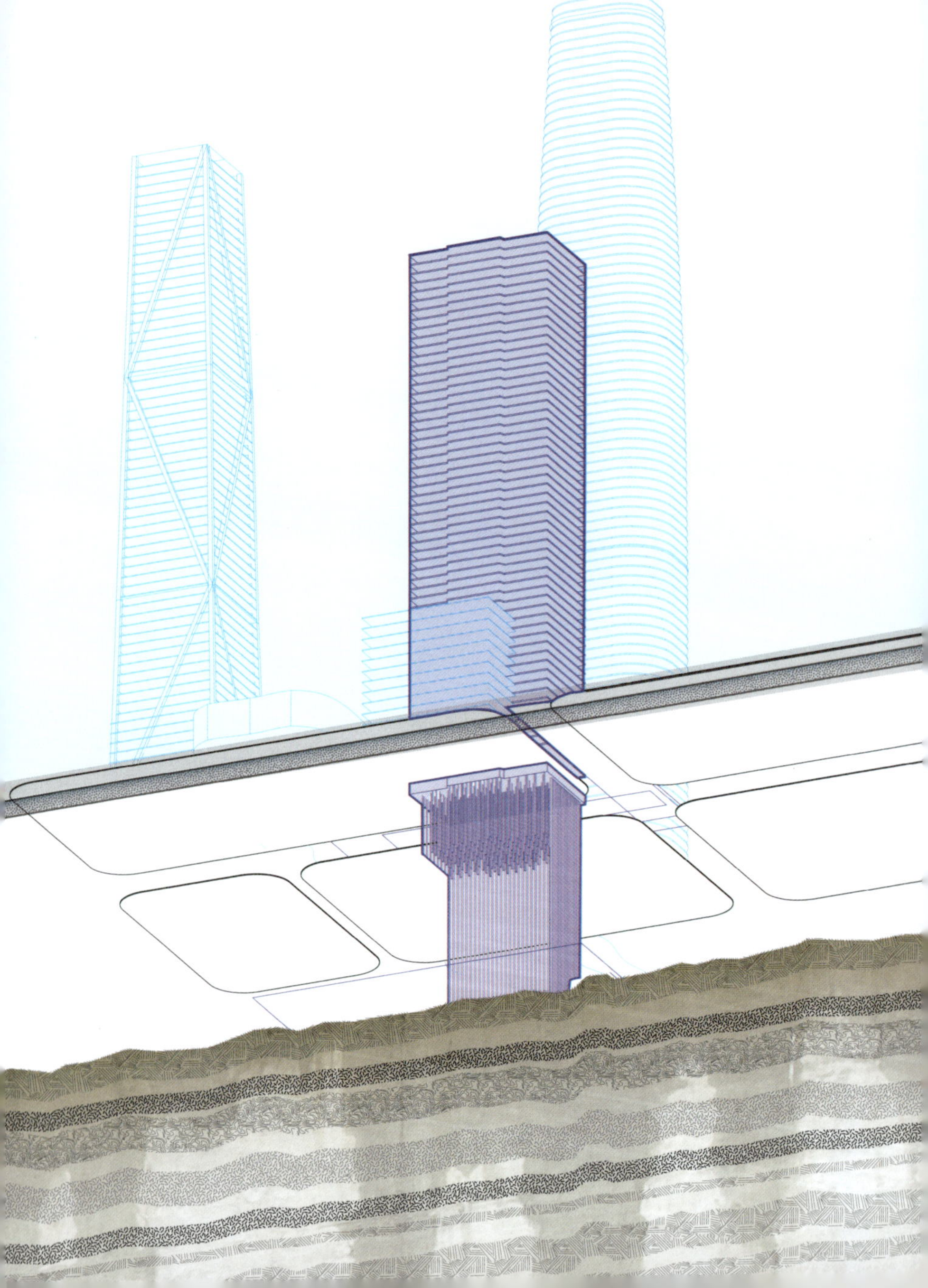

Millennium Tower

San Francisco

What is the most notorious building in San Francisco today? It does not have a particularly identifiable silhouette, and it has the somewhat generic glass-centric aesthetic of many urban towers of the early 2000s. The building is the Millennium Tower, a fifty-eight story, 645-foot-tall condominium. As of 2022, the skyscraper had sunk eighteen inches and tilted twenty-four inches to the west.[118] While the tilt has been stopped through various engineering fixes, the sinking is not unique to the 419-unit building, completed in 2009. The larger issue here is that San Francisco itself (as well as several other cities worldwide) is slowly sinking, a phenomenon known as subsidence.[119] A recent report by *Science Advances* notes that subsidence can exceed ten millimeters per year when the ground is a combination of artificial landfill and mud, as is the case in many parts of San Francisco.[120] Another contributing factor has been construction at the nearby Transbay Terminal, where up to five million gallons of water have been pumped out of the ground each month in a process known as dewatering, a crucial precursor to construction.[121] Moreover, California has gone through years of drought, further compounding the issue. For cities like San Francisco that are at risk from rising sea levels, subsidence is a dual blow.[122] And while $100 million was spent fixing the tilt at Millennium Tower—a luxury building enjoyed by wealthy residents—its woes raise a broader issue: how to address the combined problems of sinking land and rising seas, especially when they impact not just buildings but public infrastructure too.[123]

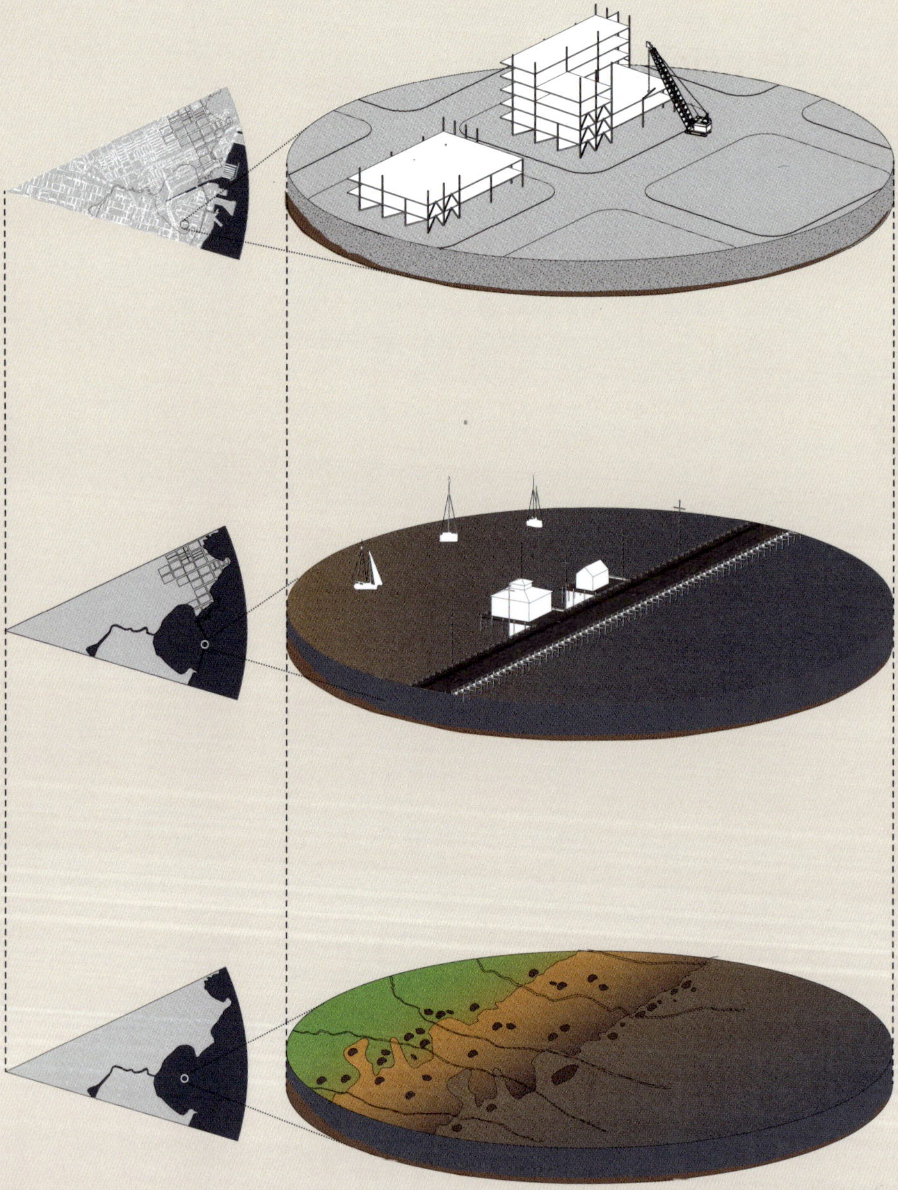

Mission Bay

San Francisco

Mission Bay, as its name suggests, used to be an estuary that combined fresh- and saltwater marshes, tidal mudflats, and a shallow bay. In this state, it was habitat for a variety of waterfowl, including ducks, geese, and herons. Originally home to the Yelamu people, it became a shipbuilding, fishing, and meat production center under European settlement in the 1800s.[124] Over the years, however, it turned into a dumping ground for debris and dilapidated ships. The waste ended up filling in Mission Bay, a pattern repeated throughout the region. During the last 150 years, 90 percent of the marshes in the San Francisco Bay have disappeared due to industrialization.[125] Once filled in, Mission Bay became home to a variety of commercial uses, including canneries, a sugar refinery, and warehouses. Recent years have brought a surge in development. As of 2024, the area is home to the Chase Center, Uber's headquarters, and the UC San Francisco Mission Bay campus. One of the lowest-lying areas of the city, Mission Bay is particularly susceptible to sea level rise. A 2016 study by the San Francisco–based think tank SPUR concluded that Mission Bay should proactively consider adaptation measures to counter rising sea levels.[126] Unlike New York, which suffered severe damage during Hurricane Sandy in October 2012, Mission Bay still has time to develop a plan before any major disaster occurs.[127] Some measures under consideration include upgrading buildings to be able to withstand flooding and raising the most flood-prone land. Perhaps Mission Bay's status as one of the lowest-lying and fastest-growing neighborhoods could be an opportunity? The area could be a testing ground for a range of experimental adaptation strategies that could then be evaluated and replicated throughout San Francisco and beyond.

Mission Creek Houseboat

San Francisco

The Mission Creek houseboats are a charming throwback to a San Francisco of another era.[128] They are a thrift-store find set against a city that increasingly defines itself through the adoption of futuristic technologies such as AI and driverless cars. But while the houseboats appear anachronistic, this area contains the complex history of many different eras. What preceded the Mission Bay neighborhood was a bay that was gradually filled in and built over.[129] An artificial waterway, now called Mission Creek, was then cut through the area. The result resembles neither the original bay nor a natural creek, instead appearing as a void in the cityscape, filled with water. The creek's present path also lies beyond the area where the tides used to terminate.[130] That area is now a convergence of swooping highways (including I-280), railways (including the northern terminus of the Caltrain commuter rail), apartment complexes, and the UC San Francisco Mission Bay campus. Amid this tapestry of development sits a community of approximately twenty houseboats of varying size, shape, and style. Some of the boats are seaworthy, while others might have trouble making it to the end of the creek. The community has existed since the 1960s, when the houseboats were relocated from nearby Islais Creek.[131] While the houseboats might seem quaint, they are also resilient. Unlike the surrounding buildings, they can and do adapt to the tides, day in and day out. They are already designed to withstand water and they can often operate off the grid and are thus not impacted by climate-related disruptions. And crucially, in a city with notoriously expensive housing, they are typically more affordable than adjacent landbound homes.[132] Could small and nimble be part of a low-carbon-footprint housing solution that also addresses adaptation to sea level rise?

drawing and AR by
Ana Panaligan and Ashley Preciado

NASA Ames Research Center

Mountain View

A facility designed to support the exploration of the far reaches of outer space for conditions suitable for human habitation is ironically under threat at ground level. The NASA Ames Research Center is one of five NASA centers at risk of flooding by 2050 due to predicted sea level rise.[133] Located in Silicon Valley, Hangar One is a two-hundred-foot-tall building covering six acres (roughly the area of six football fields). Originally built in 1933, Hangar One is one of the largest freestanding structures in the world. It was designed to house the USS *Macon*, one of the biggest naval scouting airships ever created.[134] Though the *Macon* was lost in 1935, the hangar continued to be used until the 1990s. It is now undergoing a restoration, set to wrap up in 2025, that includes the removal of toxic chemicals.[135] The site has been home to much innovative research, including efforts to shape vehicles to better withstand the extreme heat of reentry into the earth's atmosphere. But a 2014 report by the Union of Concerned Scientists included the Ames Research Center on a national list of important historical and cultural sites (one of three in California) at risk due to climate change.[136] The adjacent Moffett Federal Airfield extends to the edge of San Francisco Bay and the Stevens Creek Shoreline Nature Study Area, a site that has become integral to migrating birds along the Pacific Flyway and is under consideration for potential wetland restoration.[137] The predicament of the Ames Research Center reminds us that even as some look to outer space for potential solutions for a planet becoming less and less habitable, earthly realities require immediate attention.

drawing by Bhavi Patel

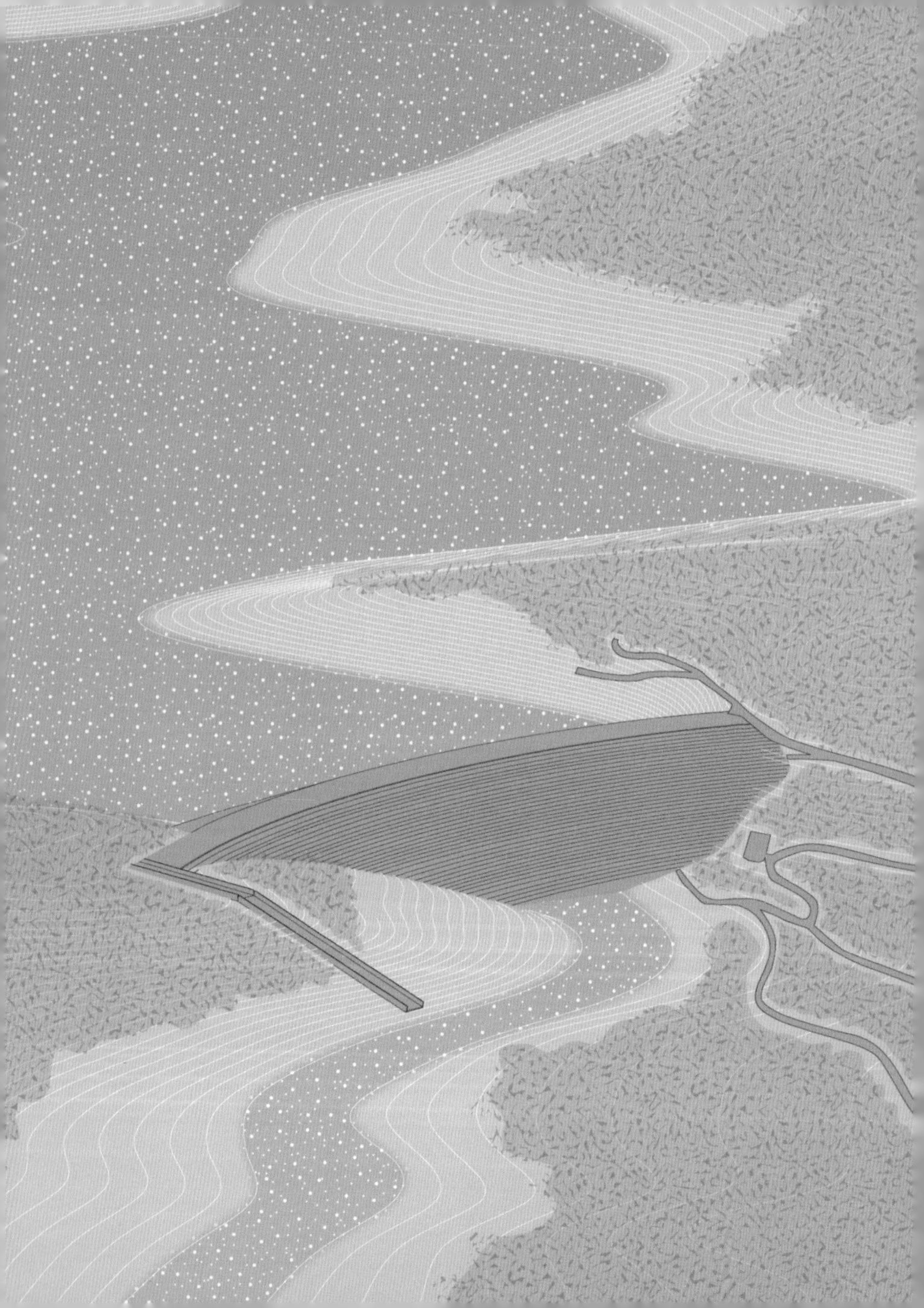

Oroville Dam

Oroville

At 770 feet high, the Oroville Dam was once considered the ninth wonder of the world; it is still the tallest dam in the United States.[138] Holding back the state's second-largest reservoir, Lake Oroville, it supplies water to 750,000 acres of farmland and twenty-seven million residents (about 70 percent of California's population).[139] The dam can provide 645 megawatts of power, about enough to power eight hundred thousand homes. But on August 5, 2021, the lake's Edward Hyatt Power Plant was turned off due to low lake levels caused by the state's historic drought. The hydroelectric plant resumed operations five months later, on January 4, 2022, following sufficient rainfall to refill the lake to normal levels.[140] The dam illustrates competing forces: On the one hand, dams provide a significant portion of California's renewable energy (from 7 percent in dry years to 20 percent in wet years).[141] On the other hand, dams can have outsize impacts on watersheds stressed by the droughts and deluges that are becoming more intense due to climate change. In 2017, for instance, the dam was compromised after heavy winter rains, resulting in the evacuation of 180,000 residents. Speaking with *Grist* magazine, Sean Turner, a hydrologist and water resources engineer, referred to potential drought related disruptions: "The impact is less likely to be power cuts and lights out, it's more likely to be increased electricity costs and potentially increased carbon emissions, because there's likely to be more reliance on gas and other resources."[142] How can we build and maintain a diverse portfolio of green energy options without causing more harm than good?

Pacifica Apartments

Pacifica

What were once apartment buildings dangerously perched on cliffs are now empty lots along Pacifica's Esplanade Avenue. Built in the 1960s, these two-story buildings featured sliding glass doors and uninterrupted ocean views. However, by 2010, severe coastal erosion had left the buildings hanging, posing a risk to occupants and beachgoers below and leading to their eventual demolition.[143] It was estimated that allowing the buildings to fall would have increased cleanup costs by ten to twenty times.[144] Why is erosion so severe here in the first place? Unfortunately, over recent decades both dams and flood control measures have reduced replenishment of the beaches by restricting sediment flow. Before dams and channelized rivers, the sediment flowing down from upstream would keep beaches healthy. Combined with rising sea levels and strong seasonal storms, erosion has intensified further. In fact, Pacifica's cliffs are some of the fastest eroding along the California coast. Studies by the Scripps Institution of Oceanography have found that some areas are disappearing at the staggering pace of up to one foot per year.[145] These increasingly dangerous conditions pose a threat to the remaining 12 percent of structures located in an at-risk coastal zone west of State Route 1, including most of the city's affordable housing and many businesses.[146] A state bill that would have supported managed retreat on a larger scale (SB 83) was passed by the California legislature but vetoed by Governor Gavin Newsom in 2021.[147] As more communities face similar choices, how can we explore creative ways for moving people and buildings out of harm's way, whether the threat is due to erosion and sea level rise or to other hazards posed by the intersection of climate change and the built environment?[148]

drawing by Stella Isaccs

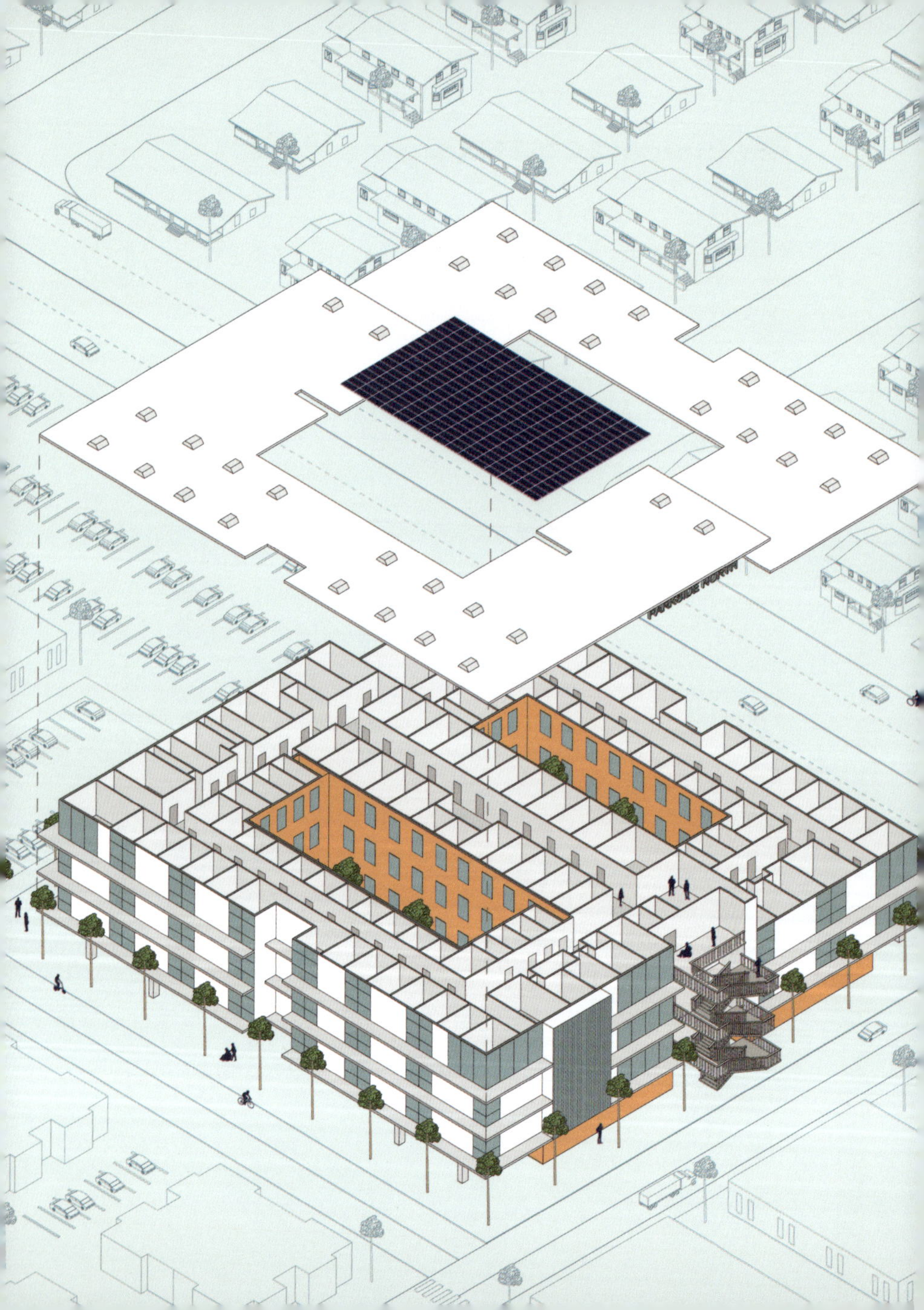
PARKSIDE NORTH

Parkside North Residence Hall

Long Beach

Could dormitories offer a sustainable model for communal living? While California laws have changed in recent years to allow increased residential density on single-family lots, dormitories fundamentally challenge assumptions about the provision of private versus shared amenities.[149] In contrast to apartments and condominiums, dormitories have a smaller ratio of private rooms to common areas. In 2021, California State University, Long Beach completed its first student housing project in thirty-four years, and it includes many examples of sustainable design. The residence hall, Parkside North, has 472 beds, pod study rooms, shared kitchens, and community areas, squeezing all of this into ninety thousand square feet, or about 191 square feet per bed. [150] Similar resources designed into single-family homes would take double or triple the square footage and require greater land usage. At the same time, Parkside North offers all of these facilities while achieving net zero energy and water usage.[151] The project was recognized both by LEED and the Living Building Challenge.[152] While LEED is a fairly well-known system intended to lower a building's carbon footprint, the Living Building Challenge goes even further, advocating for buildings that contribute positively to the environment.[153] As the state shifts away from low density single-family zoning, examples such as Parkside North look beyond what single-family structures typically provide to suggest more communal ways of living—ones that could make a serious impact on cutting greenhouse gas emissions. How might dormitories, traditionally a form of dwelling associated with students, be looked at again as a model for a broader swath of people? Could this type of housing work both in commercial districts and single-family zones?

drawing by Bhavi Patel

Platform Harmony
Santa Barbara

The offshore oil platforms begin peeking out of the ocean about one-and-a-half miles from Santa Barbara's shore. Along US Route 101, several appear in a line stretching east to west: 7 Holly, Hondo, Harmony, and Heritage, the common starting letter H tagging the platforms to their specific coastal location.[154] These are some of the nineteen that continue to sit off the coast of Santa Barbara. The collective site's infamy dates back to what remains the largest oil spill off the coast of California—4.2 million gallons released over ten days in January and February 1969.[155] The oil spill occurred at Platform A, about 5.5 miles from Santa Barbara. So much oil and gas was released that the ocean appeared to be boiling. Thousands of birds were killed during the disaster, smothered in oil from the spill, and only a lucky few were saved by volunteers.[156] Today, the most visible effects of the oil spill are long gone, but the rigs remain, creating an aesthetic dissonance between the beautiful sand beaches and California's energy needs. In response to the 1969 disaster (and about three months before the first Earth Day was celebrated), the people of Santa Barbara held Environmental Rights Day on January 28, 1970. While the rigs unwittingly helped bring about an environmental movement, however, they also continued to pump oil. But their days are numbered. In 2021, the rigs were given a ten-year decommissioning deadline by the US Bureau of Safety and Environmental Enforcement, after years of community activism led by a group called Get Oil Out, which was formed in the wake of the 1969 spill.[157] Sometimes it takes a disaster to catalyze a movement. Could reinforcing the memory of this disaster be used to propel even further action? For residents of Santa Barbara, the oil rigs are a constant reminder just off the horizon.

drawing by Stella Isaacs

Salton Sea

Salton City

In 2012, people complained of a dead fish smell coming from the Salton Sea. A decade later and most of the fish have died off and fully decomposed.[158] And no wonder: The salinity of the Salton Sea is about double that of the Pacific Ocean.[159] When conditions are windy, dried-up areas on the lake's edge release toxic dust, which has been detected as far away as Los Angeles, about 150 miles to the northwest.[160] The surrounding area's abandoned pools, houses, motels—even a yacht club—hint at a once thriving resort destination.[161] In a way, it still is an oasis: The area is a "stopover site for millions of birds of approximately 400 species," making it one of the most important migratory locations west of the Rockies.[162] The lake itself covers approximately three hundred square miles and sits 240 feet below sea level.[163] But the lake is actually an accident, formed when an irrigation canal was breached by floodwater from the Colorado River in 1905. For thousands of years, the lake would occur during wet years as it was connected to the natural flows of the Colorado River, and during dry years the sea would evaporate. Over recent decades, the Colorado River has been transformed from a natural river to a managed system that serves the needs of multiple states. Among these needs, industrial agriculture often takes priority. Consequently, less water now reaches the Salton Sea, which has lost about a third of its size.[164] Today, it is perhaps the best symbol of how water management, industrial agriculture, and drought intersect in California. As temperatures rise and water needs increase, the strain on such locations will only worsen. Perhaps the apocalyptic picture presented by the Salton Sea is best understood as both a warning of the unintended consequences of water diversion and a wake-up call for the need for sustainable water management.

San Francisco Federal Building

San Francisco

So many fences. By the entrances; around the plaza. Even the perforated metal facade resembles a giant fence enclosing the building. This level of security is probably not what the architect intended, but it has been a necessary response to the open-air drug trade that has plagued the neighborhood. The San Francisco Federal Building, designed by the Pritzker Prize–winning architect Thom Mayne and his firm Morphosis, sends mixed messages: Its angular metal exterior signals both an eco-friendly techno-futurism and a less-than-welcoming demeanor. The goal of the building, completed in 2007, was to present a repeatable model for sustainability, community, and health.[165] It indeed boasts many sustainability achievements. It takes advantage of existing conditions such as daylight (for interior lighting) and the temperate climate (for heating and cooling). Its perforated metal skin reduces direct light on its glass windows, allowing natural light without the need for mechanical air conditioning. The area between its two facades pulls hot air up, like a chimney, further lowering inside temperatures. But the fence motif—through no fault of the architect—has come to symbolize the state of the neighborhood. The grounds have a bunker-like quality, and the area has been plagued by homelessness, drug use, and post-COVID vacancies. Perhaps building management could make small adjustments to make this otherwise sustainable tower more usable. A stealthily designed skate park at United Nations Plaza (viewable from the Federal Building's Sky Lounge) has become a place where a sense of community and recreation have merged. The plaza is successful because it has a community that uses and maintains it daily. As San Francisco grapples with urban vacancy on a larger scale, the Federal Building, with its broad range of public spaces, has the potential to be part of a low-carbon solution once it finds its people.[166]

drawing and AR by
Zach Poulin and Montserrat Zamora

PORT OF
SAN FRANCISCO

San Francisco Ferry Building

San Francisco

Now a bustling destination for commuters, foodies, and tourists, the San Francisco Ferry Building, which originally opened in 1898, was once the primary gateway to the city for those arriving by a combination of transcontinental train and regional ferry.[167] It was designed in the Beaux-Arts style by architect Arthur Page Brown (who sadly did not live to see its completion) and includes a 660-foot-long skylit nave and a 245-foot-tall clock tower inspired by Spain's Seville Cathedral.[168] When it was built, skeptics worried it might fail during an earthquake or that its concrete foundation would sink into the bay. But it immediately became a San Francisco icon, gracing postcards and appearing in films like *The Maltese Falcon.*[169] When ferry transit peaked in the 1930s, some 50–60 million people crossed the bay annually on more than fifty ferry services, contributing to about 250,000 passengers flowing through the Ferry Building each day. Built prior to the Golden Gate and San Francisco-Oakland Bay Bridges, it remained largely unscathed by the 1906 earthquake that destroyed much of the city.[170] In 2006, the Embarcadero historic district was nominated to the National Register of Historic Places.[171] But in 2016, the Ferry Building was deemed one of the eleven most-endangered historic places in the United States due to seismic threats and sea level rise.[172] The area where the building sits is particularly vulnerable because the ground beneath it is no more than bay mud (with bedrock as far as 240 feet below).[173] The Port of San Francisco and the Army Corps of Engineers are considering raising the surrounding area by as much as seven feet to reduce flood risk, at a predicted cost of $13.5 billion.[174] While the area was constructed before residents knew of sea level rise, it raises questions about what gets protected and at what cost. Furthermore, as many cities convert industrial waterfronts, how might public infrastructure protect cities rather than become liabilities that need protection?

drawing and AR by
Maggie Ma and Aileen Louie

Sepulveda Adobe

Calabasas

The Sepulveda Adobe, located in Malibu State Park, was built in 1863 by Pedro Sepulveda and Soledad Sepulveda, Mexican settlers who were assisted by members of the Chumash tribe, according to Sepulveda family history. The adjacent land was filled with oak trees that provided acorns to the Chumash and Tongva people. Later, it was reportedly a source of firewood and charcoal for the Sepulvedas. It was also where their twelve children grew up.[175] The house was eventually given to the California Department of Parks and Recreation to turn into a museum. Badly damaged during a 1994 earthquake, it received seismic upgrades that were completed in 2017. Just before its public reopening, however, the house was reduced to a shell by the Woolsey Fire of 2018.[176] The fire burned 96,949 acres, destroyed 1,643 structures, caused three fatalities, and prompted the evacuation of more than 295,000 people. Total property damage was estimated at $6 billion, as reported by the California Public Utilities Commission.[177] The adobe walls of Sepulveda Adobe, a traditional Mexican building technique, were among the few elements that withstood the fire. While the direct cause of the fire was an electrical malfunction, the more diffuse cause connects to a years-long drought, the presence of invasive species in the region, development in an area prone to wildfire, and the absence of Indigenous approaches to land management, such as controlled burns. As wildfire season in California becomes more pervasive, perhaps the most crucial question is: How do we care for the places we value? This could mean a range of solutions, including burying infrastructure, creating defensible space, and removing invasive species, as well as a return to Indigenous practices like prescribed burns and the rewilding of developed land.[178] How could a futuristic vision embrace traditional land management and building techniques that have proved effective in the past?

drawing by Bhavi Patel

Solar Star

Rosamond

Six billion years. That is how long astronomers estimate we have left before the sun runs out of energy.[179] In the coming decades, solar is set to be the top energy source globally.[180] But solar's rise has been slow. In the early 1970s, the oil crisis set off research that resulted in innovations such as the experimental 1973 Solar One house developed by Karl W. Böer with solar-powered heat and energy.[181] By 1979, the Carter administration had installed solar panels on the West Wing of the White House, only to have them removed by the subsequent Reagan administration.[182] But fits and starts have gradually led to the adoption of solar energy. In 2015, Solar Star became the largest solar farm in the United States, and at the time, in the world. Located in Rosamond, an unincorporated community in the Mojave Desert, Solar Star is a 579-megawatt photovoltaic power station that has 1.7 million individual panels spread across 3,200 acres of desert.[183] The panels are set back from the road behind barbed wire and laid out in rows. They tilt imperceptibly as they track the sun throughout the day.[184] The farm provides enough energy to supply about 250,000 homes. The surrounding high desert climate sees several days a year above one hundred degrees Fahrenheit and is just a short drive from Alta Wind Farm, underscoring this location's importance as both a place experiencing the effects of climate change and one that is leading the charge away from greenhouse gas emissions.[185]

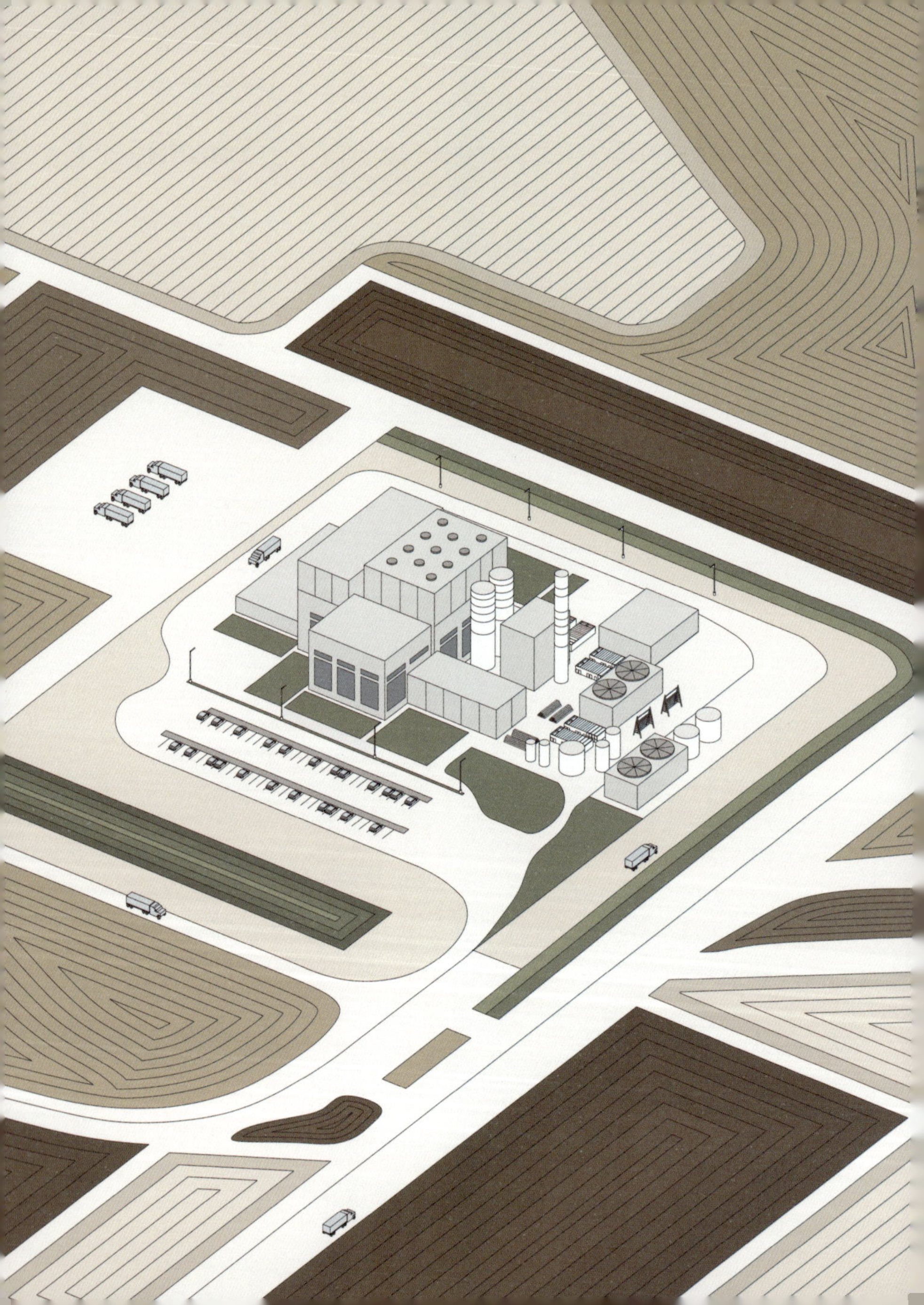

Stanislaus Resource Recovery Facility

Crows Landing

The average person in the United States generates nearly five pounds of trash per day—adding up to roughly the weight of a classic Volkswagen Beetle over a year.[186] Though most of our trash goes to landfills, a small portion ends up in waste-to-energy plants. These are places where trash is burned and the resulting heat generates electricity. Only about seventy-five such facilities operate in the United States (compared to more than five hundred in Europe).[187] While waste-to-energy may seem like a win/win, closer scrutiny reveals more of a mixed bag. As Earthjustice reports, this type of waste management facility is an impediment to a zero-waste future. They can be costly to maintain, and they are also often located in lower-income communities, exacerbating environmental injustice.[188] Despite downsides, however, they can have benefits. When managed properly, they can be a clean energy source, divert waste from landfills (reducing methane emissions), and help support the recycling of materials.[189] Stanislaus Resource Recovery Facility—now known as Reworld—illustrates these contradictions. The building, a big box with chimneys and power lines, is about an hour's drive from San Francisco and sits atop rolling hills far from residential communities. Inside, one can see a mountain of household trash. Look closer and you can spot the hallmarks of everyday life: used milk cartons, takeout containers, old shoes. Deeper inside the facility, an incinerator burns the waste into a sea of flames. The heat generated powers a turbine, and the resulting energy is sold to Pacific Gas and Electric. But as of 2024, Reworld is teetering on closure, so now is a perfect time to consider its role in the larger climate narrative.[190] As much as 1.8 million acres are taken up by landfills in the United States.[191] That is about twice the area of Rhode Island. Perhaps incinerators can provide a needed bridge until we, as a society, learn to use less and recycle more?[192]

drawing by Bhavi Patel

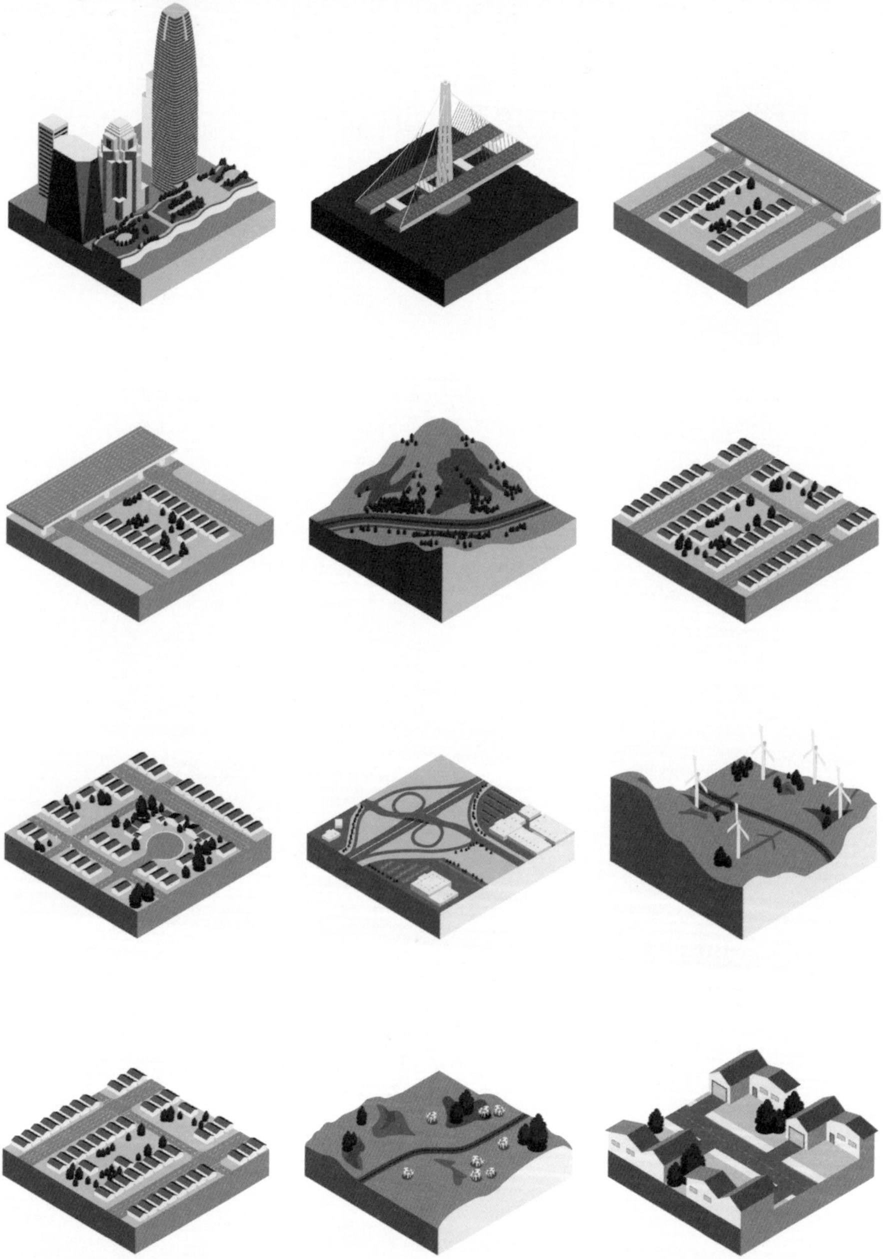

Stockton Super Commute

Stockton

As housing prices have soared in major metropolitan areas like San Francisco, people are forced to live farther from the places that are often a source of high-paying jobs. This displacement has given rise to what has become known as the "super commuter"—that is, someone with a commute time of ninety minutes or more. One city that exemplifies this trend is Stockton, where 10 percent of the workforce are super commuters.[193] In 2024, the median home price in San Francisco was about triple that of Stockton.[194] The phenomenon of super commuting is often interpreted as signaling a fundamental misalignment of jobs and affordable housing, with less attention given to the environmental impact of this misalignment. In fact, transportation accounts for almost one-third of greenhouse gas emissions in the United States.[195] Seen this way, ensuring the proximity of jobs to affordable housing is a climate issue. But how do we change this situation? One way is to offer more public transportation options to the periphery; the other is to create more affordable housing in close proximity to urban centers.[196] Though housing advocates and the state have worked to create more housing and lower costs (via legislative action such as the California HOME Act, SB 9, for example), more needs to be done.[197] How can we reflect on the successes (and failures) of recent measures to make urban centers more affordable? For architects, this could mean a variety of solutions: from repurposing underused offices for housing, to splitting properties, to developing new prefabricated housing.[198] What are the implications if we think of housing affordability as a climate issue?

drawing and AR by
Adam Overmeer and Langston Hay

Sutter's Mill

Coloma

Sutter's Mill is one of California's most famous landmarks. Located in the foothills of the Sierra Nevada on the South Fork of the American River, it is where a carpenter named James W. Marshall first discovered gold in 1848.[199] Perhaps more than any other place, this mill was the catalyst for the state's transformation over the subsequent two centuries. While early prospectors panned for gold, the work soon became industrialized, using a process known as hydraulic mining.[200] How does this work? A dam is created at high elevations, and water from the resulting reservoir is channeled through iron pipes. The water is then shot out of cannons, altering the original riverbed and sending vast amounts of sediment downstream, where the gold can be gathered. Hydraulic mining added as much as three feet of sediment from the American River into San Francisco Bay. In response, communities downstream built levees to prevent flooding from further changing the course of the river and altering habitats. Moreover, mercury was used to separate the gold from crushed rock, contaminating downstream waters. By the end of the nineteenth century, as much as eight million pounds of mercury had been released into the San Francisco Bay through this process.[201] When mercury is released into a marine ecosystem, it is first absorbed by marine plants and eaten by small organisms. Eventually it makes its way up the food chain until it is found in fish such as halibut, striped bass, and sturgeon. In certain parts of San Francisco Bay, these fish are especially prone to high mercury levels, another example of how Gold Rush–era practices shaped environmental conditions in ways that persist to this day.[202] As Bay Area communities grapple with imminent sea level rise and consider building sea walls at the ocean's edge, should they also look upstream to coordinate strategies of restoration and repair?

drawing by Langston Hay

The Theme Building

Los Angeles

The Theme Building is not really the culprit here—but it has become an icon for how Californians (and the United States more broadly) are disproportionately contributing to climate change through flying. A single international flight can emit more carbon than the average person emits in many parts of the world in a year.[203] And Los Angeles International Airport (LAX) is the busiest airport in California, serving around seventy-five million passengers per year, with an average of one plane taking off every minute.[204] Analysis of data from the website Airport Tracker (airporttracker.org) reveals that the combined emissions of LAX and other major airports—including London's Heathrow Airport, Dubai International Airport, Beijing Capital International Airport, and New York's John F. Kennedy International Airport—are as high as 72.4 million tons of CO_2 each year, the equivalent of eighteen coal plants.[205] While the Theme Building might serve as a futuristic symbol for LAX, it perhaps better represents an anachronistic vision centered around cars and planes and reliant on fossil fuels. Its presence is just as notable as the absence of a functional public transit system to access the airport.[206] Designed by Pereira & Luckman and completed in 1961, the building has itself been transformed over the years: Originally it was a marker for LAX, then it became a restaurant. As of 2024, it is being operated by the United Service Organizations and is not accessible to visitors.[207] On the horizon, Los Angeles will host the 2028 Olympics and has set a deadline of 2045 to use only renewable energy at LAX.[208] But this goal is only for the airport proper, not for the planes taking off daily. Perhaps, as we wait for more sustainable modes of air travel, we might consider flying less and marveling more at gravity-defying structures like the Theme Building.[209]

drawing by Stella Isaacs

Tulare Lake Farms

Corcoran

Located in the San Joaquin Valley, Tulare Lake was once the largest lake west of the Mississippi.[210] For centuries, it was a vital resource to the Yokut people, who called it Pa'ashi (Big Water) and built their lives around its seasonal swelling.[211] In the second half of the nineteenth century, diversion of water for municipal and agricultural purposes caused the lake to disappear, only to periodically reemerge in wet years. During this time, the Tulare Basin was thoroughly transformed into agricultural land.[212] This was enabled by a vast network of irrigation canals and the pumping of groundwater with hundreds of wells, some as deep as 3,500 feet below ground.[213] This massive irrigation infrastructure has allowed the area to become some of the most productive farmland in the country. The broader Central Valley represents just 1 percent of the farmland in the United States but produces 25 percent of the nation's food, including 40 percent of our fruit and nuts.[214] In the spring of 2023, the Tulare Basin flooded, with water stretching as far as the eye could see.[215] About ninety-four thousand acres of industrial farmland were submerged, including orchards and row crops. The flooding was caused by snowmelt flowing from the Sierra Nevada. Intensifying the situation, the basin is lower than it once was due to subsidence (as groundwater has been pumped out, the ground has sunk). Since the 1920s, some areas have sunk as much as thirty feet.[216] What the reemergence of the lake perhaps best represents is the persistence and intensification of nature's rhythms as climate change creates even greater swings between drought and deluge.[217] Though Tulare Lake may have disappeared from maps, its geographical status as the lowest point in the region never changed. In fact, it got only lower.[218] We don't usually think of groundwater as an extractive resource, but Tulare Lake's return—and the toll it took on submerged farms—reminds us that it comes at a cost.

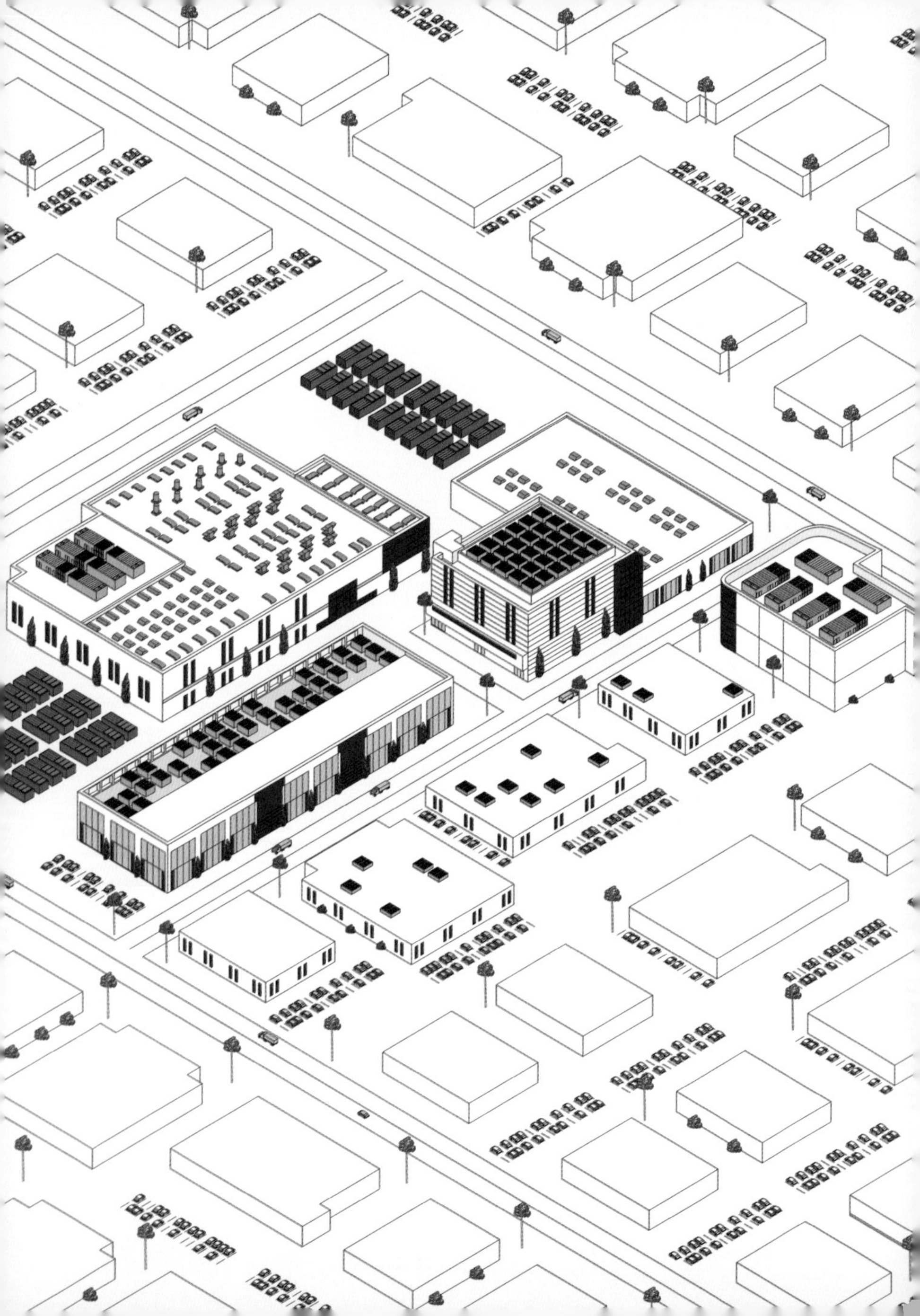

Vantage Data Center

Santa Clara

From the outside, it could be a mall, warehouse, or prison: three big boxes surrounded by a parking lot in the heart of Silicon Valley.[219] These mostly windowless facades look generic, but behind them is the state-of-the-art physical infrastructure that runs our online lives.[220] Two major advances over the last decade—the move toward data storage "in the cloud" and the shift to artificial intelligence (AI)—have massively increased the energy demands on data centers. An AI query consumes about ten times the amount of power as a conventional Google search. And research by Goldman Sachs predicts a 160 percent increase in energy demands from AI by 2030.[221] One company leading efforts to meet these demands is Vantage Data Servers, which operates more than thirty data centers worldwide, including its Santa Clara campus. Inside a Vantage building, one finds all the physical infrastructure the cloud is built on: servers, cooling equipment, generators, and security, all with layers of redundancy. While the company's Santa Clara location claims clean energy goals, including a commitment to reach net-zero operational carbon emissions by 2030,[222] commitment is not the same as actual performance. Additionally, the site is situated in a Federal Emergency Management Association–designated special flood zone, which speaks to the ways that technology can intersect with real-world risks.[223] Though the flood risk might seem low—just a 1 percent yearly chance—over a thirty-year period the chance jumps to one in four.[223] As AI factors into planning for the next generation of data centers, new criteria will likely be developed for where these buildings are erected. In fact, if you ask ChatGPT what issues are involved in planning a new data center, it lists climate risk as the number one concern. Centers are already cropping up in places like Iceland, where natural ventilation is feasible and risks from climate change are low.[224] How can we ensure that the environmental footprint of our technologies does not cloud the fundamental needs of the planet?

drawing by Bhavi Patel

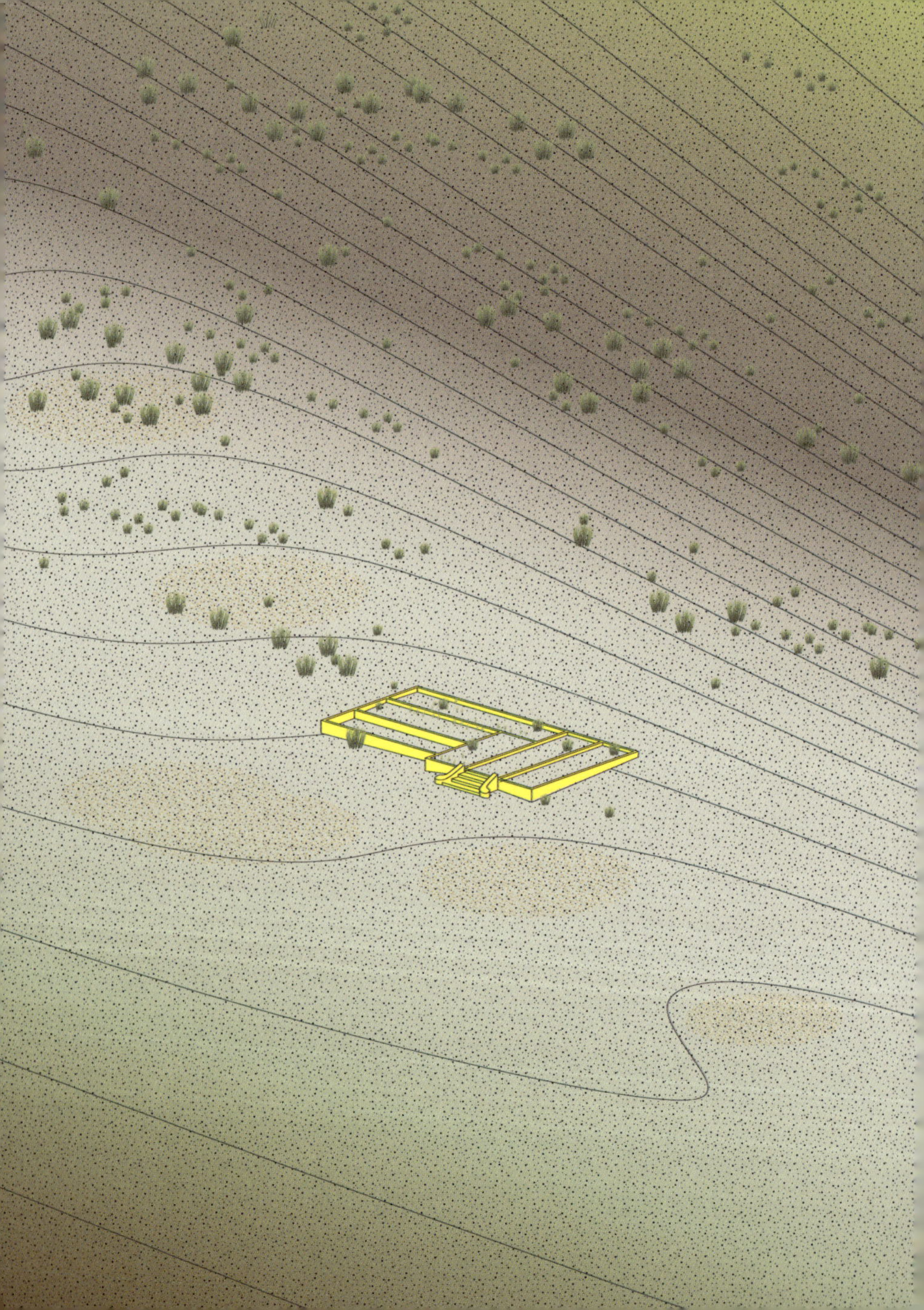

Whiskey Flat

Kernville

In September 2022, after years of drought, the long-sunken town of Whiskey Flat peeked out from Lake Isabella.[225] Why did the town disappear in the first place? In 1953, the US Army Corps of Engineers intentionally created Lake Isabella by diverting the Kern River, inundating this Gold Rush–era ghost town.[226] Though many buildings were relocated, some remained. The goal was to create flood protection, a source for irrigation, and a place for recreation. As drought lowered the waters of Lake Isabella, what emerged was a story many would have preferred to leave in the past. With the lake at 8 percent capacity, the remains of a bridge, a schoolhouse, and a church emerged, generating renewed public interest in the history of the site.[226] The town was originally developed in 1860 when gold was discovered by Lovely Rogers. Later that year, pioneer Adam Hamilton opened a saloon, initially just a plank laid over two whiskey barrels, naming it Whiskey Flat.[227] The saloon catered to prospectors and put the town on the map. The fame turned to infamy, however, when settlers clashed repeatedly with Native tribes, including the Tübatulabal and Kawaiisu peoples who had lived in the area for centuries. During a two-year period, sixty settlers and two hundred Native Americans were killed. Tensions escalated further when army Captain Moses A. McLaughlin and seventy soldiers captured one thousand Native Americans and subjected them to a two-hundred-mile Trail of Tears. In an effort to suppress this violent history, the town was renamed Kernville in 1864.[227] The creation of Lake Isabella buried the story deeper until its unexpected emergence due to drought conditions. Although the lake began to refill in 2023, once again submerging the town, its troubled history could not be buried.[228] As more places experience the dance between deluge and drought, how might stories from the past inform land management decisions for the future?

Wilshire Grand Center

Los Angeles

It has been referred to as "the grand pour." It required 2,120 truckloads and 18.5 hours of nonstop pouring of concrete. Some walls are up to four feet thick.[229] Guinness World Records cites it as the largest continuous mat pour (a mat is a foundation that holds the full load of the structure above). With twenty-one thousand cubic yards of concrete, it surpassed the Venetian Resort Las Vegas, the previous record holder.[230] To fulfill the order, eight concrete plants and a backup were all contracted within a twenty-mile radius to ensure the concrete would arrive while it was still wet. At 1,100 feet, the Wilshire Grand Center is the tallest building in Los Angeles and west of the Mississippi River. Inside the Center is a mixed-use development with retail, a shopping complex, office space, and observation decks. An express elevator whisks visitors to the seventieth floor, where they can take a second elevator to the seventy-third-level rooftop bar. The concrete that enables it all is the world's most ubiquitous building material (and the second-most-consumed material after water). But it is also what *The Guardian* calls "the most destructive material on Earth."[231] Concrete contributes as much as 8 percent of global CO_2 emissions. The world uses about thirty billion tons of concrete annually, which amounts to about four cubic feet per person per year.[232] That is the equivalent in weight to about one GMC Hummer per person. Could we imagine a requirement for designers of any large building to consider materials that are less carbon intensive? While concrete may be impossible to eliminate entirely in the foreseeable future (especially for foundations), smaller towers are already using structural alternatives like cross-laminated timber (for example, the Ascent MKE building in Milwaukee) and structural stone (for example, 15 Clerkenwell Close in London). Could some of our oldest building materials offer a less carbon-intensive path forward?

drawing by Langston Hay and Brett Snyder

Yurok Tribal Lands

Orick

Often, land acknowledgments (paying tribute to Indigenous lands at the beginning of an event) are words without tangible actions. This is a different story. In Humboldt County, long-standing efforts to reclaim land are bringing about necessary environmental healing Prior to the Gold Rush, the Yurok people were stewards of five hundred thousand acres, including 'O Rew on Prairie Creek.[233] Many of the Yurok practices of land management were and still are models of sustainability.[234] These include supporting the function of rivers and floodplains, maintaining habitat for wildlife such as salmon and beavers, and implementing prescribed burns that rejuvenate the forest.[235] During the nineteenth-century gold rush, extraction started to destroy the ecological function of these lands, as well as the culture and health of the Yurok people. Excessive logging made the forest floor susceptible to erosion, leaving a legacy of fine sediment in riverbeds and clogging gravel areas where salmon spawn. The growth of industrial-scale canneries led to overfishing, diminishing the salmon population.[236] In recent decades, Native American tribes, including the Yuroks, have implemented measures to rejuvenate their ancestral lands that will have wide-ranging effects. In 2011, the tribe obtained loans to purchase land with the idea that revenue from sustainable logging could then be used to repay the debt. Two years later, the tribe used California's Compliance Offset Program to maintain the forest without logging, selling carbon credits to companies in exchange for monitoring the area.[237] The return of some of these lands to tribal management is key to a more holistic approach to sustainability. How might projects like this—grounded in the Land Back movement and guided by Indigenous stewardship—transform exploitative and extractive land use practices into a future rooted in restoration and repair?

Afterword

I teach at UC Davis, and this project started with a simple list of places throughout California where climate change plays a key role. I thought of these places as sites of inquiry for students in a course titled Graphitecture. The class uses experimental drawing methods as a springboard to look critically at the built environment. After teaching this iteration of the course twice, I was given the opportunity to present the work in the form of an exhibition during San Francisco Design Week. Watching the audience take in the material—visitors would look slowly and consecutively at each poster (as well as the text and AR content)—spurred me to rethink the work as a book that expanded on the exhibition. I settled on fifty sites—a symbolically significant figure—and adopted a research process similar to the one I had asked my students to use. Some of the work from that class appears here in its original form. For the additional sites, I chose to focus on a representative range of issues that could make climate change tangible at a human scale.

drawing and AR by
Leah Sarber and Joelle Tahta

If the Graphitecture class is what led directly to this book, several other past collaborations also helped form the broad ideas. One such project—#OurChangingClimate with Claire Napawan and Sheryl-Ann Simpson—was premised on the idea that we can see and experience climate change around us in our everyday lives. We developed workshops to assist participants in telling those stories. One of the sites we researched for #OurChangingClimate—Mission Bay—is also included in this book and was at the center of a climate change–focused walking tour that Napawan and I led as part of San Francisco Design Week in 2017.

Another inspiration was the Resilient by Design Challenge, in which I participated as part of the Public Sediment team, led by SCAPE. This project looked at a single watershed's impact in and around San Francisco Bay: Alameda Creek. Though the site is not included in this book, what I learned from colleagues on the project was fundamental to this publication. Around the same time, collaborations with Susannah Sayler and Edward Morris also inspired a continued look at how extractive practices have fundamentally altered landscapes in California and beyond.

drawing and AR by
Nadya Verick and Elias Britton

One last project that inspired this book—and is somewhat of a companion piece—is Museum of the Phantom City, an iPhone app developed with my partner, Irene Cheng, that allows users to see visionary but unbuilt architecture on its intended site. That project, from 2009 and supported by the Van Alen Institute, used what was then a novel technology—the iPhone—to understand the built environment by comparing the world that is to the world that could be.

Unlike Museum of the Phantom City, all of the projects described in this book exist and can be seen in real life. My hope is that readers will not only recognize the sites should they visit them but will also imagine other sites that are part of this larger climate narrative. I also hope that projects featured here, such as the Klamath River dam removal project, will inspire the next generation of designers and planners to consider ways that the built environment could be rethought through a lens of healing and repair. This is as much about rethinking what should be built as what should be unbuilt.

Endnotes

Aliso Canyon

1. "Study: 2015 California Blowout Led to Largest U.S. Methane Release Ever," National Oceanic and Atmospheric Administration, February 25, 2016, https://www.noaa.gov/media-release/study-2015-california-blowout-led-to-largest-us-methane-release-ever.

2. Byron Chan, "The Lessons We Didn't Learn from the Largest Gas Leak in U.S. History," Earthjustice, October 25, 2018, https://earthjustice.org/article/the-lessons-we-didn-t-learn-from-the-largest-gas-leak-in-u-s-history.

3. Amina Khan, "Porter Ranch Leak Declared Largest Methane Leak in U.S. History," *Los Angeles Times,* February 26, 2016.

4. "Importance of Methane," US Environmental Protection Agency, last updated October 21, 2024, https://www.epa.gov/gmi/importance-methane.

5. Byron Chan, "Report: Utility's Negligence to Blame for Los Angeles Methane Disaster," Earthjustice, June 20, 2019, https://earthjustice.org/experts/byron-chan/report-utility-s-negligence-to-blame-for-los-angeles-methane-disaster.

6. "Porter Ranch Residents Protest against Aliso Canyon Gas Facility," news report, August 22, 2023, posted August 23, 2023, by FOX 11 Los Angeles, YouTube, 2:22, https://youtu.be/i_9u6DNSD3s.

7. Tianjun Lu et al., "Leveraging Citizen Science and Low-Cost Sensors to Characterize Air Pollution Exposure of Disadvantaged Communities in Southern California," *International Journal of Environmental Research and Public Health* 19, no. 14 (2022): 8777.

Alta Wind Energy Center

8. Mari Nøvik and Jennifer Hiller, "Wind Industry in Crisis as Problems Mount; More than $30 Billion in Spending Is Delayed as Setbacks Pile Up for Renewable-Energy Sector," *Wall Street Journal,* August 7, 2023.

9. "Wind Vision," US Department of Energy, Office of Energy Efficiency and Renewable Energy, https://www.energy.gov/eere/wind/wind-vision.

10. For more about distributed wind energy, see "Distributed Wind Energy," US Department of Energy, Office of Energy Efficiency and Renewable Energy, WINDExchange, https://windexchange.energy.gov/markets/distributed.

11. Heather Doty, "Small-Scale Distributed Wind Projects Could Leave a Large Impact," US Department of Energy, Office of Energy Efficiency and Renewable Energy, July 17, 2024, https://www.energy.gov/eere/wind/articles/small-scale-distributed-wind-projects-could-leave-larg.

Badwater Weather Station

12. "Death Valley Hits 130 Degrees as Heat Wave Sweeps the West," *New York Times,* July 10, 2021.

13. Amy Graff, "The Fight over Death Valley's 134-Degree Temperature Record Heats Up," *San Francisco Gate,* July 24, 2023.

14. "NPS and NWS Announce New Weather Station at Badwater Basin," press release, September 8, 2021, National Park Service, https://www.nps.gov/deva/learn/news/badwater-weather-station.htm.

15. Concepción de León and John Schwartz, "Death Valley Just Recorded the Hottest Temperature on Earth," *New York Times,* August 17, 2020.

16. Katharine Gammon, "'Like an Oven': Heat and Tourism Create Headache for US Park Rangers," *Guardian,* July 28, 2023.

Balboa Island House

17. "History of Balboa Island," Visit Newport Beach California, https://visitnewportbeach.com/history-of-balboa-island/.

18. Scott Martelle, "Out Here; Rising Tides at Balboa Island," *Los Angeles Times,* December 23, 2018.

19. Rosanna Xia, "California Against the Sea," *Los Angeles Times,* July 7, 2019.

Bateson Building

20. Simon Sadler, "The Bateson Building, Sacramento, California, 1977–81, and the Design of a New Age State," *Journal of the Society of Architectural Historians* 75, no. 4 (2016): 469–89.

21. Patricia Leigh Brown, "It Happened Here First," *New York Times,* November 17, 2005.

Capistrano Beach Boardwalk

22. Anh Do, "Residents Fear Losing Capistrano Beach as Storms Chip Away at Their 'Piece of Paradise,'" *Los Angeles Times,* December 7, 2018.

23. "Coastal Crisis: California's Vanishing Beaches," originally aired July 11, 2022, on KLCS-TV, posted July 12, 2022, by OC World, YouTube, 29:30, https://youtu.be/NysuKaCqS18.

24. Jacob Aere, "Full Passenger Rail Service Between San Diego and Orange County Expected to Resume in Early April," KPBS, March 11, 2024, https://www.kpbs.org/news/quality-of-life/2024/03/11/full-passenger-rail-service-between-san-diego-and-orange-county-expected-to-resume-in-early-april.

25. Bring Back Our Beaches, https://www.bringbackourbeaches.com/.

26. "Save Capistrano Beach!," Surfrider Foundation, November 16, 2022, https://southoc.surfrider.org/campaigns/Save Capistrano Beach!.

27. Natalie Tso, "California's Beaches Are Eroding: Here's How to Save Them," UCI Samueli School of Engineering, https://engineering.uci.edu/news/2023/6/californias-beaches-are-eroding-heres-how-save-them.

Carlsbad Desalination Plant

28. Carlsbad Desal Plant, https://www.carlsbaddesal.com/.

29. Tony Perry, "Backers of Desalination Hope Carlsbad Plant Will Disarm Critics," *Los Angeles Times,* June 4, 2015.

30. George Courser, "The REAL Truth Behind Carlsbad 'Bud' Lewis Poseidon Desalination Plant: A Costly Mistake," *Southern Sierran* (blog), March 7, 2022, Sierra Club Angeles Chapter, https://angeles.sierraclub.org/news_conservation/blog/2022/03/the_real_truth_behind_carlsbad_bud_lewis_poseidon_desalination_plant.

31. Karen Lykkebo Petersen et al., "Biological and Physical Effects of Brine Discharge from the Carlsbad Desalination Plant and Implications for Future Desalination Plant Constructions," *Water* 11, no. 2 (2019): 208.

Cherokee Studios

32. Annie Block, "They've Got Rhythm: Los Angeles' Cherokee Studios Mixed-Use Lofts," *Interior Design,* June 1, 2010, https://interiordesign.net/projects/they-ve-got-rhythm-los-angeles-cherokee-studios-mixed-use-lofts/.

33. "Cherokee Studios West Hollywood, CA," Brooks + Scarpa, https://brooksscarpa.com/cherokee-studios.

34. John Linden, "AIA COTE 2011 Top Ten Green Projects: Cherokee Studio," *Architect,* https://www.architectmagazine.com/photos/aia-cote-2011-top-ten-green-projects-cherokee-studio.

Church of One Tree

35. Maci Martell, "Photos: Santa Rosa's Church of One Tree, Built 150 Years Ago, a Local Marvel," *Santa Rosa Press Democrat,* September 21, 2023.

36. Andrew C. Isenberg, *Mining California: An Ecological History* (Hill and Wang, 2006).

37. Isenberg, *Mining California.*

38. Trevor Nace, "Best Way to Fight Climate Change? Clone 3,000-Year-Old Redwoods," *Forbes,* July 29, 2019.

39. "Protect the Clar Tree," Guerneville Forest Coalition, https://www.guernevilleforestcoalition.org/the-clar-tree.

40. Garrison Frost, "Helping Redwoods Fight Climate Change," *Redwoods Magazine,* June 20, 2023, Save the Redwoods League, https://www.savetheredwoods.org/redwoods-magazine/helping-redwoods-fight-climate-change/.

Corcoran Bowl

41. Lois Henry and Ryan Christopher Jones, "The Central California Town That Keeps Sinking," *New York Times,* May 25, 2021.

42. Susanne Rust, Jessica Garrison, and Ian James, "The Farm Barons of the Tulare Lake Basin Want to Continue Drawing Groundwater at Volumes That Are Collapsing the San Joaquin Valley," *Los Angeles Times,* December 31, 2023.

43. Hayley Smith, "Levee Project to Get State Boost; Newsom Pledges Millions to Help Protect Corcoran from Rising Waters," *Los Angeles Times,* May 13, 2023.

Donald C. Tillman Water Reclamation Plant

44. "A Postcard from the Field: Los Angeles Sanitation's Japanese Gardens | Dateline NBC," posted November 20, 2019, by Dateline NBC, YouTube, 2:46, https://www.youtube.com/watch?v=M0147Xh4eko.

45. David Lustig, "A Botanical Fantasy in Van Nuys," *Los Angeles Times,* Valley Edition, May 28, 1993.

46. Evelyn De Wolfe, "Japanese Garden Grows from Reclaimed Water," *Los Angeles Times,* May 16, 1993.

Dutch Slough

47. Tara Duggan, "A Gigantic Appetite for Greenhouse Gas," *San Francisco Chronicle*, April 26, 2023.

48. "Dutch Slough Tidal Habitat Restoration 2021," YouTube video, 4:41, posted by "California DWR," November 23, 2021, https://www.youtube.com/watch?v=qG-8nbYZ6ro.

49. Duggan, "A Gigantic Appetite."

Fassler House

50. Greg Fassler, interview with author, March 6, 2023.

51. "Replanting Pacifica State Beach," *Pacifica Tribune*, March 31, 2004.

52. Chris Hunter, "Fassler House Demolished for Beach Improvement," *Pacifica Tribune*, October 1, 2003.

Foster City Levee

53. John King, "Foster City's Bet on Sea Level Safety," *San Francisco Chronicle*, November 22, 2023.

54. "Foster City: A Story of Filling the Bay," *San Francisco Gate*, June 14, 2009, https://www.sfgate.com/bayarea/article/foster-city-a-story-of-filling-the-bay-3295665.php.

55. Katie Spidalieri and Annie Bennett, "Managed Retreat Toolkit," Georgetown Climate Center, https://www.georgetownclimate.org/adaptation/toolkits/managed-retreat-toolkit/introduction.html?chapter.

Fresno High-Speed Rail

56. John Donegan, "Renderings Come with a Reminder: A Bullet Train Is Still Years Away from Bakersfield," *Bakersfield Californian*, April 30, 2024.

57. "Carbon Footprint Calculator Text-Only Version," California High Speed Rail Authority, https://hsr.ca.gov/programs/green-practices-sustainability/carbon-footprint-calculator/carbon-footprint-calculator-text-only-version/.

58. Stacy C. Davis and Robert G. Boundy, *Transportation Energy Data Book: Edition 39*, ORNL/TM-2020/1770 (Oak Ridge National Laboratory, 2021), https://tedb.ornl.gov/wp-content/uploads/2021/02/TEDB_Ed_39.pdf.

59. California High-Speed Rail Authority, https://hsr.ca.gov/.

Harris Ranch

60. "About Us," Harris Ranch Beef Company, https://www.harrisranchbeef.com/about-us/.

61. "Stop: Fresno-Coalinga Rd.," Invisible-5, https://www.invisible5.org/index.php?page=fresnocoalinga

62. "Agriculture and Aquaculture: Food for Thought," US Environmental Protection Agency, Southeast New England Program, October 2020, last updated February 6, 2024, https://www.epa.gov/snep/agriculture-and-aquaculture-food-thought.

63. Naoki Nitta, "Why Can't We Just Quit Cows?," *Grist*, November 1, 2023, https://grist.org/agriculture/why-cant-we-just-quit-cows/.

64. "Methane: A Crucial Opportunity in the Climate Fight," Environmental Defense Fund, https://www.edf.org/climate/methane-crucial-opportunity-climate-fight.

Healdsburg Memorial Bridge

65. Mark S. Hoddle, "Giant Reed," UC Riverside, Center for Invasive Species Research, https://cisr.ucr.edu/invasive-species/giant-reed.

66. Hoddle, "Giant Reed."

67. Kate Golden, "On the Russian River, a Slow Road to Good Fire," Bay Nature, May 30, 2024, https://baynature.org/article/on-the-russian-river-a-slow-road-to-good-fire/.

68. Jana Lawyer, "Arundo Removal Project Receives Matching Grant!," Russian Riverkeeper, July 9, 2023, last updated 2024, https://russianriverkeeper.org/arundo-removal/.

Heirloom Carbon Technologies

69. Brad Plumer, "In a U.S. First, a Commercial Plant Starts Pulling Carbon from the Air," *New York Times*, November 9, 2023.

70. Plumer, "In a U.S. First."

71. Brad Plumer, "A Huge City Polluter? Buildings. Here's a Surprising Fix," *New York Times*, March 10, 2023.

Inglewood Oil Field

72. Alan Taylor, "The Urban Oil Fields of Los Angeles," *The Atlantic*, August 26, 2014.

73. Alexandria Herr and Clayton Aldern, "California's Dirty Little Secret: Oil Wells in the Backyard," *Grist*, October 13, 2001, https://grist.

org/project/accountability/wilmington-california-oil-gas-setbacks/.

74. Phil Willon, "New California Oil Drilling Must Be Set Back from Homes and Schools, Newsom Says," *Los Angeles Times,* October 21, 2021.

Integral Urban House

75. Farallones Institute, *The Integral Urban House: Self-Reliant Living in the City* (Random House, 1982).

76. Helga Olkowski and Bill Olkowski, *The City People's Book of Raising Food* (CreateSpace Independent Publishing, 2014).

77. Farallones Institute, *The Integral Urban House.*

78. Greg Castillo, "Hippie Modernism," *Places,* October 2015, https://doi.org/10.22269/151026.

Kaufmann Desert House

79. Simon Niedenthal, "'Glamourized Houses': Neutra, Photography, and the Kaufmann House," *Journal of Architectural Education* 47, no. 2 (1993): 101–12.

80. "About Us," Modernism Week, https://modernismweek.com/pages/about-us.

81. Amy Dvorak, "How Neutra's Kaufmann House Got Its Groove Back," *Dwell,* March 13, 2019, https://www.dwell.com/article/kaufmann-house-richard-neutra-8520af09.

82. Kendall Balchan, "Report Details What Climate Change May Mean for Palm Springs, Surrounding Area," *Palm Springs Post,* June 30, 2022.

83. Salvador Hernandez, Joseph Serna, and Liam Dillon, "Intense Heat Wave Brings New California Fires; Palm Springs Breaks All-Time Heat Record," *Los Angeles Times,* July 5, 2024.

84. Soumya Karlamangla and Maggie Miles, "Palm Springs Is Heating Up, and Only Getting Hotter," *New York Times,* July 18, 2023.

Kellogg House

85. "Rich Bar (No. 337 California Historical Landmark)," Sierra Nevada Geotourism, https://sierranevadageotourism.org/entries/rich-bar-no337-california-historical-landmark/2043aad4-84b9-4b20-b097-d51de8afb693.

86. Lauren Hernández, "Dixie Fire Tears Through Historic Town of Greenville in Plumas County," *San Francisco Chronicle,* August 4, 2021.

87. "Rebuild | Reburn; Wildfires' Grim Reality; Rural Climate Skeptics Are Costing Us Time and Money. Do We Keep Indulging Them?," *Los Angeles Times,* October 2, 2022.

88. Stuart Thornton, "After the Gold Rush," *National Geographic,* last updated June 26, 2024, https://education.nationalgeographic.org/resource/after-gold-rush.

Klamath River Dam Removal

89. Greg Fuhs, "Dams Targeted for Removal in Effort To Restore Rivers," *Water Environment and Technology* 12, no. 1 (2000): 22–23.

90. Ann Willis, "Dam Removal on the Klamath: Reflections on How We Got Here," American Rivers, January 11, 2024, https://www.americanrivers.org/2024/01/dam-removal-on-the-klamath-reflections-on-how-we-got-here/.

91. Ian James, "Restoring Life to an Ancient Watershed; Reservoirs Are Drained as Huge Dam Removal Project Advances on the Klamath River," *Los Angeles Times,* March 26, 2024.

Laguna Beach Goats

92. "Laguna Beach Goat Vegetation Management Project," FEMA, last updated February 11, 2021, https://www.fema.gov/case-study/laguna-beach-goat-vegetation-management-project.

93. Russell A. Beatty, "Prescribed Grazing," *Landscape Architecture* 95, no. 3 (2005): 50–57.

94. Karolina Rivas, "Over 1,000 Goats Needed to Clear Dry Brush from California to Help Prevent Wildfires," ABC News, October 18, 2019, https://abcnews.go.com/US/1000-goats-needed-clear-dry-brush-california-prevent/story?id=66368853.

95. Clara Germani, "Want to Prevent Wildfires? Try Getting Someone's Goat," *Christian Science Monitor,* November 19, 2020, https://www.csmonitor.com/USA/Society/2020/1104/Want-to-prevent-wildfires-Try-getting-someone-s-goat.

Lancaster Solar

96. Felicity Barringer, "With Help from Nature, a Town Aims to Be a Solar Capital," *New York Times,* April 8, 2013.

97. Barringer, "With Help from Nature."

98. "Lancaster Prison," Baker Electric, https://www.baker-electric.com/projects/lancaster-prison.

99. "How Much Solar Energy Do Homes Produce?," USAFacts, last updated November 9, 2023, https://usafacts.org/articles/how-much-solar-energy-do-homes-produce/.

100. Amy Thomson, "California Just Became the First State to Require Solar Panels on New Residential Construction," *Mother Jones,* May 9, 2018, https://www.motherjones.com/environment/2018/05/california-just-became-the-first-state-to-require-solar-panels-on-new-residential-construction/.

101. "Lancaster Mayor Honored as 'Green Mayor of the Year' for Work in Energy, Sustainability," PR Newswire, May 15, 2024, https://www.prnewswire.com/news-releases/lancaster-mayor-honored-as-green-mayor-of-the-year-for-work-in-energy-sustainability-302146694.html.

Los Angeles River

102. Michael Kimmelman, "Remaking the River That Remade L.A.," *New York Times Magazine,* November 10, 2022, https://www.nytimes.com/interactive/2022/11/10/magazine/la-river-redesign.html.

103. "Reimagining the Los Angeles River," Earth Focus, season 5, episode1, directed by Matthew Crotty, originally aired on PBS, April 3, 2024, posted April 3, 2024, by PBS SoCal, YouTube, 26:40 https://www.youtube.com/watch?v=6T7hlh_q4Ok.

104. "Lewis MacAdams: L.A. River Activist and Poet," PBS SoCal, January 21, 2011, https://www.pbssocal.org/shows/departures/lewis-macadams-l-a-river-activist-and-poet.

105. Linh Tat, "L.A. Unveils Its Vision for Sepulveda Basin," *Los Angeles Daily News,* June 18, 2024.

Marine Cloud Brightening

106. Anthony Edwards, "Bay Area Hosts Global Warming Experiment," *San Francisco Chronicle,* April 7, 2024.

107. Christopher Flavelle and Ian C. Bates, "Warming Is Getting Worse. So They Just Tested a Way to Deflect the Sun," *New York Times,* April 2, 2024.

108. Caelyn Pender, "Alameda City Council Unanimously Blocks Climate Change Project from Resuming on USS Hornet," *Mercury News,* June 4, 2024.

109. Edwards, "Bay Area Hosts Global Warming Experiment."

110. Elizabeth Kolbert, "Climate Change from A to Z," *New Yorker,* November 21, 2022.

Mighty Buildings

111. "The Carbon Challenge," Carbon Leadership Forum, https://carbonleadershipforum.org/the-carbon-challenge/.

112. "Fast Facts on Transportation Greenhouse Gas Emissions," US Environmental Protection Agency, last updated June 18, 2024, https://www.epa.gov/greenvehicles/fast-facts-transportation-greenhouse-gas-emissions.

113. Mighty Buildings, https://www.mightybuildings.com/.

114. "Case Study Houses," Los Angeles Conservancy, https://www.laconservancy.org/save-places/issues/case-study-houses/.

115. Jane Margolies, "Energy-Efficient Isn't Enough, So Homes Go 'Net Zero,'" *New York Times,* November 16, 2021.

116. "Sustainable Management of Construction and Demolition Materials," US Environmental Protection Agency, last updated January 23, 2024, https://www.epa.gov/smm/sustainable-management-construction-and-demolition-materials.

117. Eleanor Gibson, "Rael San Fratello 3D Prints Mud Structures as Prototypes for Low-Cost Construction," *Dezeen,* October 3, 2019, https://www.dezeen.com/2019/10/03/mud-frontiers-rael-san-fratello-3d-printed-low-cost-construction/.

Millennium Tower

118. Ricardo Cano, "S.F. Tower Draws New Residents Despite Tilt," *San Francisco Chronicle,* June 5, 2022.

119. Linda Poon, "Cities Are Sinking Under the Weight of Urban Development," Bloomberg, February 23, 2021, https://www.bloomberg.com/news/articles/2021-02-23/big-buildings-can-cause-cities-to-sink-faster.

120. Manoochehr Shirzaei and Roland Bürgmann, "Global Climate Change and Local Land Subsidence Exacerbate Inundation Risk to the San Francisco Bay Area," *Science Advances* 4, no. 3 (2018): eaap9234, https://doi.org/10.1126/sciadv.aap9234.

121. J. K. Dineen, "Agency Says It Will Keep Pumping Water Near Leaning SF Tower," *San Francisco Gate,* September 22, 2016.

122. Shirzaei and Bürgmann, "Global Climate Change."

123. Kasha Patel, "To Keep from Sinking, Cities Put Water Back into Ground," *Washington Post,* June 8, 2023.

Mission Bay

124. Gary Kamiya, "Bay Marshlands Lure 1st Residents," *San Francisco Chronicle,* January 23, 2016.

125. John King, "Along San Francisco's Mission Creek, Sea Level Rise Unsettles the Waters, *San Francisco Chronicle,* June 4, 2021, https://www.sfchronicle.com/projects/2021/san-francisco bay-area-sea-level-rise-2021/mission-creek.

126. *Mission Creek Sea Level Rise Adaptation Study: Waterfront Strategies for Long Term Urban Resiliency* (SPUR, September 2016), https://www.spur.org/publications/research/2016-09-26/mission-creek-sea-level-rise-adaptation-study.

127. N. Claire Napawan and Brett Snyder, "Porosity: Networking Cities for a Changing Climate," *Architecture_MPS* 6, no. 1 (2014), https://doi.org/10.14324/111.444.amps.2014v6i1.001.

Mission Creek Houseboat

128. Carl Nolte, "S.F.'s Plucky Floating Community Watches Waves of Change," *San Francisco Gate,* March 2, 2014.

129. Nancy Olmsted, "Mission Bay Gazeteer [sic] of Historic Places," FoundSF, https://www.foundsf.org/index.php?title=Mission_Bay_Gazeteer_of_Historic_Places.

130. For those exploring the site, the intersection of Division and King Streets is where the original creek emptied into the bay.

131. Nolte, "S.F.'s Plucky Floating Community."

132. "Floating Residents: Couple Stays Afloat in Mission Creek Home | The Regulars," Jessica Christian, YouTube video, 3:33, posted August 3, ,2021, by San Francisco Chronicle, YouTube, 3:33, https://www.youtube.com/watch?v=XoFf7qk_C8Y.

NASA Ames Research Center

133. Debra Holtz et al., National Landmarks at Risk: How Rising Seas, *Floods, and Wildfires Are Threatening the United States' Most Cherished Historic Sites* (Union of Concerned Scientists, May 2014), https://www.ucsusa.org/resources/national-landmarks-risk.

134. "Hangar One, Mountain View, California," Atlas Obscura, July 10, 2015, http://www.atlasobscura.com/places/hangar-one.

135. Rachel Hoover, "Hangar One Restoration Project," NASA, April 25, 2022, last updated July 26, 2023, https://www.nasa.gov/history/hangar-one-restoration-project/.

136. Holtz et al., *National Landmarks at Risk.*

137. "What Is NASA Doing to Protect Its Facilities from Sea-Level Rise?," NASA Sea Level Change, https://sealevel.nasa.gov/faq/22/what-is-nasa-doing-to-protect-its-facilities-from-sea-level-rise.

Oroville Dam

138. Don Sweeney, "Bigger than a Breadbox? How Oroville Dam Stacks Up," *Sacramento Bee,* February 15, 2017.

139. "Oroville Dam," Water Education Foundation, https://www.watereducation.org/aquapedia-background/oroville-dam.

140. "California DWR Resumes Operations at 645 MW Hyatt Powerplant," *Hydro Review* (blog]), January 6, 2022, https://www.hydroreview.com/business-finance/business/california-dwr-resumes-operation-at-645-mw-hyatt-powerplant/.

141. Alvar Escriva-Bou, Gokce Sencan, and Andrew Ayres, "California's Water and Energy Systems Are Inextricably Linked," PPIC blog, December 6, 2022, https://www.ppic.org/blog/californias-water-and-energy-systems-are-inextricably-linked/.

142. Emily Pontecorvo, "The West's Biggest Source of Renewable Energy Depends on Water. Will It Survive the Drought?," *Grist,* October 18, 2022, https://grist.org/drought/will-western-hydropower-survive-drought/.

Pacifica Apartments

143. Lyanne Melendez, "Crews Demolish Condemned Pacifica Apartment Complex," ABC7 News, January 30, 2017, https://abc7news.com/pacifica-rain-damage-coastal-access-path crews-install-fence-around-beach-swallowed-by-sinkhole/1728963/.

144. Bay City News Service, "Pacifica: Erosion Sets Stage for Coastal Apartments to Be Demolished," *Alameda Times-Star,* March 7, 2016.

145. Steve Koppes, "New High-Resolution Study on California Coastal Cliff Erosion Released," Scripps Institution of Oceanography, August 4, 2022, https://scripps.ucsd.edu/news/new-high-resolution-study-california-coastal-cliff-erosion-released.

146. Grace Scullion, "New Study Highlights Speed of Pacifica's Erosion," *Pacifica Tribune,* August 16, 2022.

147. "California Senate Bill 83," LegiScan, https://legiscan.com/CA/text/SB83/id/2799070.

148. Jake Bittle, "As California Attempts a 'Managed Retreat,' Coastal Homeowners Sue to Stay," *Grist,* May 10, 2023, https://grist.org/housing/california-managed-retreat-half-moon-bay-coastal-commission/.

Parkside North Residence Hall

149. Hannah Wiley, "A New California Housing Law Has Done Little to Encourage Building, Report Says," *Los Angeles Times,* January 18, 2023.

150. "Parkside Village," Cal State Long Beach, University Housing and Residential Life, https://www.csulb.edu/student-affairs/university-housing-residential-life/parkside-village.

151. David Malone, "California State University Long Beach's New Dormitory Is One of California's Most Sustainable," *Building Design + Construction,* August 9, 2021, https://www.bdcnetwork.com/california-state-university-long-beachs-new-dormitory-one-californias-most-sustainable.

152. Dawn Hammon, "New LEED-Targeted Student Housing Supports Net-Zero Goals," Inhabitat, August 27, 2021, https://inhabitat.com/new-leed-platinum-student-housing-supports-net-zero-goals/.

153. "Living Building Challenge," Living Future, https://living-future.org/lbc/.

Platform Harmony

154. "Drivers Can/t Miss View of Oil Platforms," *Lompoc Record,* May 6, 2008.

155. Roman Mars, host, 99% Invisible, episode 338, "Crude Habitat," produced by Emmett FitzGerald, January 22, 2019, https://99percentinvisible.org/episode/crude-habitat/.

156. "The Spill That Energised US 'Green' Movement," posted June 6, 2010, by Al Jazeera English, YouTube, 2:40 https://www.youtube.com/watch?v=nYPTdLB1C-o.

157. Jean Yamamura, "Say So Long to Santa Barbara's Offshore Platforms," *Santa Barbara Independent,* January 19, 2022, https://www.independent.com/2022/01/19/so-long-offshore-platforms/.

Salton Sea

158. Saige Miller, "The Salton Sea Shows Why Utah Should Pay Attention to Great Salt Lake's Stench," KUER 90.1, January 2, 2024, https://www.kuer.org/health-science-environment/2024-01-02/the-salton-sea-shows-why-utah-should-pay-attention-to-great-salt-lakes-stench.

159. Michael Cohen, "Salton Sea +20," Pacific Institute, September 13, 2019, https://pacinst.org/salton-sea-20/.

160. Hector Becerra, "Area's Odor Is Linked to Salton Sea; Air Quality Officials Say a Storm Blew the Smell 150 Miles into the L.A. Basin," *Los Angeles Times,* September 12, 2012.

161. Madeline Hiltz, "Salton City: From Desert Oasis to Abandoned Wasteland," Abandoned Spaces, February 1, 2022, https://www.abandonedspaces.com/towns/salton-city-desert-wasteland.html.

162. Alecia Smith, "A Shorebird Lover's Paradise at the Salton Sea," Audubon California, May 29, 2024, https://ca.audubon.org/news/shorebird-lovers-paradise-salton-sea.

163. Henry Fountain and Mette Lampcov, "The Salton Sea, an Accident of History, Faces a New Water Crisis," *New York Times,* February 25, 2023.

164. Jules Bernstein, "Why the Salton Sea Is Turning into Toxic Dust," UC Riverside News, July 27, 2024, https://news.ucr.edu/articles/2022/10/06/why-salton-sea-turning-toxic-dust.

San Francisco Federal Building

165. "San Francisco Federal Building," Morphosis, https://www.morphosis.com/architecture/12/.

166. J. K. Dineen, "Mission Street in Midst of Major Revitalization." *San Francisco Chronicle,* July 17, 2014.

San Francisco Ferry Building

167. "A Brief History of Ferries on the Bay . . . ," *Bay Crossings* (blog), January 2001, http://baycrossings.org/dispnews.php?id=582.

168. "Ferry Building: San Francisco, California,"

American Planning Association, https://www.planning.org/greatplaces/spaces/2010/ferrybuilding.htm.

169. John King, "Perfect Symbol of S.F., Ferry Building Is 125," *San Francisco Chronicle*, July 11, 2023.

170. "A Brief History of Ferries."

171. "Embarcadero Historic District," Port of San Francisco, https://sfport.com/node/6543.

172. John King, "Embarcadero in SF on List of Nation's at-Risk Historic Treasures," *San Francisco Gate*, October 5, 2016.

173. John King, "Quakes, Floods Forecast to Ruin S.F. Embarcadero," *San Francisco Chronicle*, September 19, 2020.

174. Joe Kukura, "Plan to 'Lift' the Ferry Building by Seven Feet Raises Alarm of Ferry Building Operator," SFist, April 4, 2024, https://sfist.com/2024/04/04/plan-to-lift-the-ferry-building-by-seven-feet-raises-alarm-of-ferry-building-operator/.

Sepulveda Adobe

175. Christine Vicars, "Saving a Part of the Past: Volunteers Take Lead in Restoring 125-Year-Old Sepulveda Adobe," *Los Angeles Times*, Valley Edition, January 24, 1988.

176. Jennifer Medina, "See How the Fires Burned Where California Goes to Escape It All," *New York Times*, November 23, 2018.

177. California Public Utilities Commission, Safety and Enforcement Division, Electrical Safety and Reliability Branch, *Investigation Report of the Woolsey Fire* (California Public Utilities Commission, 2018), https://www.cpuc.ca.gov/-/media/cpuc-website/divisions/safety-and-enforcement-division/investigations-wildfires/sed-investigation-report-woolsey-fire---redacted.pdf.

178. "Defensible Space," CAL FIRE, https://www.fire.ca.gov/dspace.

Solar Star

179. "Chapter 6: Aging into Gianthood," in "Life and Death of a Planetary System," NASA Science, last updated. April 2024, https://science.nasa.gov/exoplanets/resources/life-and-death/chapter-6/.

180. "Massive Expansion of Renewable Power Opens Door to Achieving Global Tripling Goal Set at COP28," International Energy Agency, January 11, 2024, https://www.iea.org/news/massive-expansion-of-renewable-power-opens-door-to-achieving-global-tripling-goal-set-at-cop28.

181. UDaily Staff, "In Memoriam: Karl W. Böer," *UDaily*, April 23, 2018, https://www.udel.edu/udaily/2018/april/in-memoriam-karl-w-boer/.

182. "White House Solar Panel," Smithsonian, National Museum of American History, https://americanhistory.si.edu/collections/nmah_1356218.

183. Cabe Atwell, "The Biggest Utility-Scale Solar Farms on Earth," ElectronicDesign, March 14, 2019, https://www.electronicdesign.com/technologies/power/article/21807720/the-biggest-utility-scale-solar-farms-on-earth.

184. Scott Becker, "A Look Into America's Largest Solar Farm," Solstice, March 22, 2019, https://solstice.us/solstice-blog/a-look-into-americas-largest-solar-farm/.

185. Declan Butler, "Solar Power: California's Latest Gold Rush," *Nature* 450, no. 7171 (2007): 768–69.

Stanislaus Resource Recovery Facility

186. "Municipal Solid Waste," US Environmental Protection Agency, last updated March 29, 2016, https://archive.epa.gov/epawaste/nonhaz/municipal/web/html/index.html.

187. "Energy Recovery from the Combustion of Municipal Solid Waste (MSW)," US Environmental Protection Agency, last updated January 30, 2024, https://www.epa.gov/smm/energy-recovery-combustion-municipal-solid-waste-msw.

188. "Vestiges of Environmental Racism: Closing California's Last Two Municipal Waste Incinerators," Earthjustice, November 9, 2021, https://earthjustice.org/feature/california-municipal-waste-incinerators.

189. Helena Nageler-Petritz, "The State of Waste-to-Energy in the US," WMW by ISWA, May 10, 2023, https://waste-management-world.com/resource-use/the-state-of-waste-to-energy-in-the-us/.

190. Ken Carlson, "Garbage-Burning Plant Plagued by Financial Issues," *Modesto Bee*, February 2, 2024.

191. Kayla Vasarhelyi, "The Hidden Damage of Landfills," University of Colorado Boulder, Environmental Center, Division of Student Affairs, April 15, 2021, https://www.colorado.edu/ecenter/2021/04/15/hidden-damage-landfills.

192. Kait Siegel, "How Our Trash Contributes to Climate Change—And What We Can Do About It," Clean Air Task Force, September 30, 2022, https://www.catf.us/2022/09/how-our-trash-contributes-to-climate-change/.

Stockton Super Commute

193. Marisa Kendall, "'Super Commutes' Become the Norm: Grueling Bay Area Drives Are Getting Worse, but One City Has It Particularly Bad," *East Bay Times*, April 25, 2018.

194. "California Housing Market," Redfin, https://www.redfin.com/state/California/housing-market.

195. "Sources of Greenhouse Gas Emissions," US Environmental Protection Agency, last updated October 22, 2024, https://www.epa.gov/ghgemissions/sources-greenhouse-gas-emissions.

196. Tom Li, "Central Valley-Bay Area Super Commuters May See a Public Transit Solution as Early as 2027," *Mercury News*, July 17, 2024.

197. "SB 9: The California H.O.M.E. Act," California Senate, https://focus.senate.ca.gov/sb9.

198. "San Francisco Announces a New Initiative to Spur Conversions of Underutilized Office Buildings," press release, June 22, 2023, SF.gov, https://www.sf.gov/news/san-francisco-announces-new-initiative-spur-conversions-underutilized-office-buildings.

Sutter's Mill

199. History.com Editors, "California Gold Rush," History, April 6, 2010, last updated August 10, 2022, https://www.history.com/topics/19th-century/gold-rush-of-1849.

200. Andrew C. Isenberg, *Mining California: An Ecological History* (Hill and Wang, 2006)

201. "Saving the Bay—The Aftermath of the Gold Rush: Mining and Mercury in San Francisco Bay," posted August 11, 2010, by KQED, YouTube, 4:39, https://www.youtube.com/watch?v=PV547grgDL4.

202. Joe Eaton, "Bay Fish Still Not Good Eating," *Estuary News* 30, no. 2 (June 2021): 3–4.

The Theme Building

203. Niko Kommenda, "How Your Flight Emits as Much CO_2 as Many People Do in a Year," *The Guardian*, July 19, 2019, https://www.theguardian.com/environment/ng-interactive/2019/jul/19/carbon-calculator-how-taking-one-flight-emits-as-much-as-many-people-do-in-a-year.

204. "10-Year Summary of Passengers," Los Angeles World Airports, https://www.lawa.org/lawa-investor-relations/statistics-for-lax/10-year-summary/passengers.

205. Sam Pickard and Ipek Gençsü, *The Airport Tracker: Uncovering Aviation Emissions*, ODI Policy Brief (ODI, 2021), https://media.odi.org/documents/ODI_Policy_Airport_Tracker_v8.pdf.

206. Colleen Shalby, "LAX's People Mover Is Likely Delayed Until 2025; Credit Agency Cites Strained Relationship Between Airport and Project's Contractor," *Los Angeles Times*, March 30, 2024.

207. "Theme Building, LAX," Los Angeles Conservancy, https://www.laconservancy.org/learn/historic-places/theme-building-lax/.

208. "LAWA Details Sustainability Progress in Its Latest Annual Report," press release, August 10, 2023, Los Angeles World Airports, https://www.lawa.org/news-releases/2023/news-release-032.

209. James Bikales and Hannah Sampson, "America's Airport Glow-Up," *Washington Post*, December 11, 2022.

Tulare Lake Farms

210. Shawn Hubler and Mark Abramson, "A Vast Lake Has Captivated California Where Farms Stood a Year Ago," *New York Times*, June 25, 2023.

211. Ian James, "Body of Water, Spirit Reawakened; Tachi Yokut Tribe Calls for the Preservation of California's Tulare Lake, Saying Its Return Should Help Heal an Ecosystem Drained by Agriculture," *Los Angeles Times*, June 27, 2023.

212. Hubler and Abramson, "A Vast Lake Has Captivated California."

213. "Central Valley Aquifer System," in "Ground Water Atlas of the United States: California, Nevada, HA 730-B," US Geological Survey, https://pubs.usgs.gov/ha/ha730/ch_b/B-text3.html.

214. "The Central Valley: Sacramento Valley," US Geological Survey, California Water Science Center, https://ca.water.usgs.gov/projects/central-valley/sacramento-valley.html.

215. Hubler and Abramson, "A Vast Lake Has Captivated California."

216. Susanne Rust, Jessica Garrison, and Ian James, "The Farm Barons of the Tulare Lake Basin Want to Continue Drawing Groundwater at Volumes That Are Collapsing the San Joaquin Valley," Los Angeles Times, December 31, 2023

217. Ian James, "Big Rain Year Has Limited Effect on Groundwater; Levels Are Up in Many Parts of California, but Aquifers Remain Low After Decades of Heavy Pumping," *Los Angeles Times,* October 16, 2023.

218. "Tulare Basin," *Lay of the Land,* no. 47 (Winter 2024): 1, 17–26, https://clui.org/newsletter/winter-2024-47/tulare-basin.

Vantage Data Center

219. "Vantage Data Centers—Santa Clara Campus," Yposted May 27, 2014, by Vantage DataCenters, YouTube, 3:39, https://www.youtube.com/watch?v=r9XVtn-Bnls.

220. "Data Sheets," Vantage Data Centers, https://vantage-dc.com/resources/datasheets/.

221. "AI Is Poised to Drive 160% Increase in Data Center Power Demand," Goldman Sachs, May 14, 2024, https://www.goldmansachs.com/insights/articles/AI-poised-to-drive-160-increase-in-power-demand.

222. "Sustainable by Design," Vantage Data Centers, April 25, 2020, https://vantage-dc.com/features/sustainability/.

223. "Are You in a Flood Zone?," Valley Water, https://www.valleywater.org/floodzone.

224. Fiona Jackson, "How Environmentally Friendly Are Iceland's Data Centers?—Part 3," TechHQ, December 1, 2023, https://techhq.com/2023/12/how-environmentally-friendly-are-icelands-data-centers-part-3/.

Whiskey Flat

225. Ronny Reyes, "California Megadrought Reveals Hidden Ghost Town Whiskey Flat," *Daily Mail,* September 9, 2022, https://www.dailymail.co.uk/news/article-11189539/California-megadrought-reveals-notorious-hidden-ghost-town-Whiskey-Flat.html.

226. Cecilia Rasmussen, "L.A. Then and Now; The Ghost of a Town Emerges from the Past; The Ruins of Old Kernville, Long Buried Under the Waters of Lake Isabella, Are Visible Again as Years of Drought Shrink the Reservoir," *Los Angeles Times,* October 24, 2004.

227. Rasmussen, "L.A. Then and Now."

228. Katie Dowd, "Ghost Town Disappears as California Lake Fills for First Time in Years," *San Francisco Gate,* July 16, 2023, https://www.sfgate.com/centralcoast/article/wild-west-ghost-town-disappears-california-lake-18203267.php.

Wilshire Grand Center

229. "The Wilshire Grand Center," Conco, https://www.conconow.com/portfolio/the-wilshire-grand-center/.

230. Thomas Curwen, "Grand Ambition; Starting on the Right Foot; Orchestrating Record-Setting Concrete Pour Was the Job of a Lifetime," *Los Angeles Times,* April 27, 2014.

231. Jonathan Watts, "Concrete: The Most Destructive Material on Earth," *Guardian,* February 25, 2019, https://www.theguardian.com/cities/2019/feb/25/concrete-the-most-destructive-material-on-earth.

232. "Concrete Needs to Lose Its Colossal Carbon Footprint," *Nature* 597, no. 7878 (2021): 593–94.

Yurok Tribal Lands

233. Julie Watson, "California Tribe That Lost 90% of Land During Gold Rush to Get Site to Serve as Gateway to Redwoods," CapRadio, March 20, 2024, https://www.capradio.org/197180.

234. Associated Press, "California Tribe Becomes the First to Manage Land with National Park Service," *Guardian,* March 20, 2024, https://www.theguardian.com/us-news/2024/mar/20/yurok-tribe-land-gold-rush-redwoods-national-park-service.

235. Samuel Gilbert, "Nine Practices from Native American Culture That Could Help the Environment," *Washington Post,* April 22, 2024

236. Kathleen Sloan, "Yurok and the Klamath River: Yurok Historical Context and Data for Assessing Current Conditions and the Effects of the proposed Klamath Restoration Project on Yurok Tribal Trust Assets and Yurok Resources of Cultural and Religious Significance," report prepared for the US Department of the Interior, Bureau of Indian Affairs, February 2011, https://media.fisheries.noaa.gov/dam-migration/yurok_klamath_doi_2011.pdf.

237. Carolyn Kormann, "How Carbon Trading Became a Way of Life for California's Yurok Tribe," *New Yorker,* October 10, 2018.

Acknowledgments

I am deeply grateful to my students who have agreed to share their work in this book, to colleagues who generously gave feedback, and to friends and colleagues with whom I have collaborated and who inspired this project. I am especially thankful to the four students who aided in the production of this book: Langston Hay, Stella Isaacs, Bhavi Patel, and Meloddy Robles. Their dedication was inspiring.

I am thankful to my partner Irene Cheng, who lent her critical eye at several stages. And I am thankful to my family for giving me time, space, and feedback. I am also indebted to the many colleagues who read drafts of site texts and helped shape this project throughout its development (including suggesting sites at the beginning of the project). These colleagues include Daniel Barber, Allison Brooks, Ben Busse, David de la Peña, Beth Ferguson, David Gissen, James Graham, Haven Kiers, Janette Kim, Claire Napawan, Albert Narath, Timothy Reardon, Simon Sadler, Emily Schlickman, and Bz Zhang. I also thank those who helped me at various phases of the project with advice and guidance, including Kadambari Baxi, Neeraj Bhatia, Mallika Bose, McLain Clutter, Jay Lund, Jonathan Massey, Nicholas de Monchaux, Thomas de Monchaux, Edward Morris and Susannah Sayler of the Canary Project, Joan Ockman, and Ann Willis. And I thank my colleague Julie Sze, who generously read over my initial book proposal and site list and offered supportive and constructive editorial feedback. I am also thankful to my teaching assistant, Utkarsh Drolia, who helped me at the beginning of my augmented reality journey; to Alireza Vaziri Rahimi and Tamiko Thiel for inspiring me with their own AR work; and to the team at Artivive for giving me technical assistance with the AR component of the book. Two pieces of writing were also inspirational at different points in this project: *A People's Guide to the San Francisco Bay Area* by Rachel Brahinsky and Alexander Tarr (University of California Press, 2020) and *H Is for Hope: Climate "Change from A to Z"* by Elizabeth Kolbert (Ten Speed Press, 2024). The first suggests ways to think about places that are not necessarily in the canon but deserve attention, and the second suggests ways of writing that address academic topics in a way that makes them accessible to a broad audience. I am also appreciative to Paula Linares and Dawn Zidonis for giving me the chance to exhibit this work in its early stages as part of San Francisco Design Week.

Over the period I researched sites, I visited as many as I could and spoke with people with direct knowledge of them, including Greg Fassler, who grew up in the Fassler house; Ren Brownell, for sharing her enthusiasm and deep knowledge of the Klamath dam; and Dustin Fuller and Lisa Shaw, for sharing their knowledge of the Corcoran Bowl. I am also indebted to Pat Kramer at the Pacifica Historical Society; Mark S. Hoddle, director of the Center for Invasive Species; Barbara Tejada, Ann Stansell, and Allison Frye of the California State Parks; Kelly McLaughlin of Reworld; Francine Magno with the City of Pacifica; and Mary Burke of California Trout. Many of these individuals also read drafts of texts and offered valuable feedback. I'm also grateful to my brother and sister in law, Jeremy and Bridget, for letting me camp out while I researched sites in Southern California.

My thanks also to the University of California, which not only provided funding but also gave resources, including access to local newspapers that covered many of the topics in the book. I thank my partner, Irene Cheng, as well as Stella Isaacs, Christopher Davey, and Hen Stanford for their help in editing the text. I am grateful for the legibility they have brought to the ideas. Lastly, I thank the publisher, ORO, which met this project with enthusiasm from start to finish.